Introduction to Microbiology

11ᵗʰ Hour

Introduction to Microbiology

Darralyn McCall

Naropa University
Boulder, Colorado

David Stock

Department of Biology
Stetson University
Deland, Florida

Phillip Achey

Department of Microbiology and Cell Science
University of Florida
Gainesville, Florida

b

**Blackwell
Science**

Editorial Offices:
Commerce Place, 350 Main Street, Malden, Massachusetts 02148, USA
Osney Mead, Oxford OX2 0EL, England
25 John Street, London WC1N 2BL, England
23 Ainslie Place, Edinburgh EH3 6AJ, Scotland
54 University Street, Carlton, Victoria 3053, Australia
Other Editorial Offices:
Blackwell Wissenschafts-Verlag GmbH, Kurfürstendamm 57, 10707 Berlin, Germany
Blackwell Science KK, MG Kodenmacho Building, 7-10 Kodenmacho Nihombashi, Chuo-ku, Tokyo 104, Japan

Distributors:
USA

Blackwell Science, Inc.
Commerce Place
350 Main Street
Malden, Massachusetts 02148
(Telephone orders: 800-215-1000 or 781-388-8250; fax orders: 781-388-8270)

Canada

Login Brothers Book Company
324 Saulteaux Crescent
Winnipeg, Manitoba, R3J 3T2
(Telephone orders: 204-837-2987)

Australia

Blackwell Science Pty, Ltd.
54 University Street
Carlton, Victoria 3053
(Telephone orders: 03-9347-0300; fax orders: 03-9349-3016)

Outside North America and Australia

Blackwell Science, Ltd.
c/o Marston Book Services, Ltd.
P.O. Box 269
Abingdon
Oxon OX14 4YN
England
(Telephone orders: 44-01235-465500; fax orders: 44-01235-465555)

Acquisitions: Nancy Whilton
Development: Jill Connor
Production: Irene Herlihy
Manufacturing: Lisa Flanagan
Marketing Manager: Carla Daves
Director of Marketing: Lisa Larsen
Interior design by Colour Mark
Cover design by Madison Design
Typeset by Best-set Typesetter Ltd., Hong Kong
Printed and bound by Capital City Press

Printed in the United States of America
00 01 02 03 5 4 3 2 1

Library of Congress Cataloging–in–Publication Data

McCall, Darralyn.
    Introduction to microbiology / by Darralyn McCall.
        p. cm.
    Includes bibliographical references and index.
    ISBN 0-632-04418-7
    1. Microbiology—Outlines, syllabi, etc. I. Title.

QR62.M325 2001
579—dc21   00-063020

CONTENTS

Unit III: Medical Microbiology, Immunology, Environmental and Applied Microbiology 153

11TH HOUR GUIDE TO SUCCESS

The 11th Hour Series is designed to be used when the textbook doesn't make sense, the course content is tough, or when you just want a better grade in the course. It can be used from the beginning to the end of the course for best results or when cramming for exams. Both professors teaching the course and students who have taken it have reviewed this material to make sure it does what *you* need it to do. The material flows so that the process keeps your mind actively learning. The idea is to cut through the fluff, get to what you need to know, and then help you understand it.

Essential Background. We tell you what information you already need to know to comprehend the topic. You can then review or apply the appropriate concepts to conquer the new material.

Key Points. We highlight the key points of each topic, phrasing them as questions to engage active learning. A brief explanation of the topic follows the points.

Topic Tests. We immediately follow each topic with a brief test so that the topic is reinforced. This helps you prepare for the real thing.

Answers. Answers come right after the tests; but, we take it a step farther (that reinforcement thing again), we explain the answers.

Clinical Correlation or Application. It helps immeasurably to understand academic topics when they are presented in a clinical situation or an everyday, real-world example. We provide one in every chapter.

Demonstration Problem. Some science topics involve a lot of problem solving. Where it's helpful, we demonstrate a typical problem with step-by-step explanation.

Chapter Test. For more reinforcement, there is a test at the end of every chapter that covers all of the topics. The questions are essay, multiple choice, short answer, and true/false to give you plenty of practice and a chance to reinforce the material the way you find easiest. Answers are provided after the test.

Check Your Performance. After the chapter test we provide a performance check to help you spot your weak areas. You will then know if there is something you should look at once more.

Sample Midterms and Final Exams. Practice makes perfect so we give you plenty of opportunity to practice acing those tests.

The Web. Whenever you see this symbol 🖥 the author has put something on the Web page that relates to that content. It could be a caution or a hint, an illustration or simply more explanation. You can access the appropriate page through *http://www.blackwellscience.com*. Then click on the title of this book.

The whole flow of this review guide is designed to keep you actively engaged in understanding the material. You'll get what you need fast, and you will reinforce it painlessly. Unfortunately, we can't take the exams for you!

PREFACE

A time-honored method of studying any subject is to condense the material in the textbook and lecture notes into "flash cards," an outline, or a few quick review sheets. We have condensed the material covered in this book into a concentrated form that you can use as quick review sheets for introductory microbiology. You can also use this book as a preview of the material you will be studying when you read your textbook or attend lectures. Reviewing the material in this book after you've read your textbook and lecture notes will help you retain what you've been studying. For students having difficulty understanding key topics, this module will prove to be an indispensable aid.

The book is intended as a supplement to your microbiology textbook and lecture notes, or for anyone who wants a review of microbiology. This book doesn't cover all the topics that might be part of an introductory microbiology course. Instead, it focuses on the more difficult material that requires the most study to master.

We have also provided practice tests for each chapter, as well as midterm and final exams for students to test their knowledge. The figures in the book are limited because your text has figures that will accompany the material in this book. You will find it useful to review those figures as you read this book. An asterisk (*) next to a problem or question indicates that it is a difficult question.

When you see the [web icon] in this book, it indicates there is additional material on the Internet related to this topic. The Web site is located at: *http://www.blackwellscience.com.* The site contains additional practice questions and material as well as test-taking tips.

Students who would benefit from this book include microbiology or biology majors, health science majors, or anyone enrolled in a non-major's introductory microbiology course. Others who would benefit might be studying for an exam based on such a course, such as the Medical College Admissions Test (MCAT) or a professional certification exam.

Good luck with your study. We hope you enjoy it.

ACKNOWLEDGMENTS

Bringing a book from an idea in someone's mind to the bookshelf is a team effort. As a group, we would like to thank Nancy Whilton, Jill Connor, and Amy Novit for their help and guidance. We've enjoyed working with them in the development of one of Nancy Whilton's modules. We would also like to thank the reviewers: William Lorowitz, Pittsburgh State; Jimmy Clark, Pittsburgh State; Michelle Riley, Pittsburgh State; Ann Smith, University of Maryland; Ed Nelson, Southeastern Louisiana University; Fred Quinn, Emory University/CDC; Don Lehman, University of Delaware; Bobbie Pettriess, Wichita State; Robert Smith, University of Science in Philadelphia, whose suggestions on presentation of this material has been invaluable.

The drawings in Chapters 6, 7, and 19 were done with ACD/ChemSketch v4.01 by Advanced Chemistry Development, Inc., 133 Richmond Street West, Suite 605, Toronto, Ontario M5H 2L3 Canada. Web: *http://www.acdlabs.com.*

INDIVIDUAL ACKNOWLEDGMENTS AND DEDICATIONS

From Darralyn McCall: I would like to thank my friend and mentor, Dan Caldwell for recommending me to be a member of this project team. I would also like to thank Phil Achey and Dave Stock for being great writing partners. I want to dedicate this book to my sister and brother-in-law: Andrea and Gary Presse; three special teachers, Dan Caldwell, Jean Cooper, and John Baranway; and to some special friends, The Greene family: Rebecca, Steven, Jessica, and Megan, and to the Boxer kids. Finally, I'd like to dedicate this to the memory of my grandparents, Elma and Robert McCall and Ruby and C. J. Galusha.

From David Stock: I thank my colleagues for giving me a great experience: Phil Achey for inviting me to become a part of this team and giving me a new experience and Darralyn McCall for the many hours she spent reading and commenting on my writing. I also thank my son James for being tolerant of my overtaking almost all the space in the house and for several helpful discussions we had as I was writing my parts of this work.

From Phil Achey: I acknowledge Jill Connor, Assistant Development Editor of Blackwell Science for her wise guidance and extreme patience with my sometimes snail's pace of writing. Secondly, I could not have finished this writing without the support, understanding, and encouragement of my wife, Melanie, who tolerated many lost evenings and weekends while I was at the computer. Thirdly, I am grateful for the efforts and contributions of my co-authors. Without them, this review would not have been completed in any timely manner, nor would it be as rich in content.

Darralyn McCall
David Stock
Phillip Achey

UNIT I

STRUCTURE, PHYSIOLOGY, AND CONTROL OF MICRO-ORGANISMS

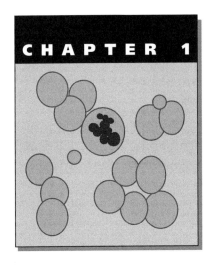

CHAPTER 1

Introduction to Microbiology

This review covers the properties of a diversity of microorganisms and their beneficial and harmful actions. The Table of Contents indicates the range of subjects reviewed. The emphasis of this review is on topics commonly contained in an introductory microbiology course and on areas that have been noticed to be difficult for students to master.

ESSENTIAL BACKGROUND

- **The scientific method**
- **Distinguish prokaryotes from eukaryotes**

TOPIC 1: THE SCIENTIFIC METHOD AND MICROBIOLOGY AS A SCIENCE

KEY POINTS

✓ *What is the scientific method?*

✓ *What organisms are studied by the microbiologist?*

✓ *What are major areas of interest to the microbiologist?*

The integrity of microbiology as a science depends on the correct application of the **scientific method**. The goal of science is to analyze available information (**observations**) and develop an explanation (**hypothesis**) for them. A hypothesis is not merely an educated guess, generalization, or prediction. An example of a prediction is that the sun will rise in the east. An example of a hypothesis is that the sun rises in the east because the earth rotates from west to east. A scientist may be asked to answer questions about observations. For example, why are some oranges sweeter than others? Why are lemons sour? Applying the scientific method to answer these questions requires developing a hypothesis to explain the answers and then confirming or refuting the hypothesis with **experimental observations**. It has been shown experimentally that the sweetness of citrus depends to a large extent on the sugar and citric acid content. When a hypothesis is contradicted by observations, it is modified so that the expected results and the observed results agree.

The scientific method requires that a hypothesis can be tested by experiment or observation. A hypothesis that has been confirmed by experiment or observation becomes a **scientific theory**.

A hypothesis that was abandoned because experiments showed it to be incorrect was the spontaneous generation of life.

When the same observation applies to many different systems (e.g., the total mass and energy of a system never changes), scientists formulate a statement called a **natural law**. A natural law describes what happens and a theory attempts to explain why it happens.

Basic science provides the foundation upon which applied science is built. A good example is the discovery, in the 1970s, that bacteria contain restriction enzymes that recognize foreign DNA and attack it. This intellectual curiosity led to the whole area of genetic engineering. Restriction enzymes play a key role in constructing specific combinations of genes that can be used in the directed change of genes in organisms. Sometimes significant discoveries are an accidental byproduct of other experimental work. A famous example is the work of William Perkin, who was attempting to synthesize quinine in the laboratory. The result of his work was a dark precipitate that stained his laboratory bench cleaning cloth a beautiful purple color. This precipitate is now known as analine dye.

The **classical definition of microbiology** is the study of living organisms that cannot be seen by the naked eye. This works very well for viruses, bacteria, and protozoa. Algae and fungi are frequently considered to be microorganisms, even though they have life cycles that may include forms visible to the naked eye. Bacteria are usually cast in the view of being harmful to our well-being. Nothing could be further from the truth. Greater than 95% of bacteria are beneficial and essential to our lives as we know them. Those that are harmful get a "larger press."

Topic Test 1: The Scientific Method and Microbiology as a Science

True/False

1. A scientific hypothesis is similar to a prediction.

2. A scientific hypothesis becomes a scientific theory only when it is supported by observations.

Multiple Choice

3. Which (if any) of the following applies to the scientific method?
   a. A hypothesis results from the experimental verification of a theory.
   b. Theories develop from hypotheses.
   c. Neither of the above apply
   d. Both of the above apply

4. Microorganisms
   a. Include only bacteria and viruses.
   b. Are usually harmful to our well-being.
   c. Are merely an intellectual curiosity.
   d. Do not play an essential role in our environment.
   e. Are considered to be visible with the naked eye.
   f. None of the above

Short Answer

5. A famous quote of Pasteur is the following: "Chance favors the prepared mind." What did he mean by this statement?

6. The scientific method requires an underlying order to nature. Einstein said: "God is subtle but He is not malicious." Explain in the context of the scientific method.

Topic Test 1: Answers

1. **False.** A scientific hypothesis is more than a prediction. It also explains why an observation occurs.

2. **True.** Observation and experiment are essential steps in the scientific method.

3. **c.** Theories are formed from hypotheses.

4. **f.** All the other choices are incorrect descriptions.

5. Unexpected experimental results might be discarded before considering possible meaning. An example was contamination of bacterial plates by fungi, observed by Fleming to result in inhibition of bacterial growth. Others before Fleming merely discarded the contaminated plates. Fleming followed up on the observation and discovered antibiotic production by the contaminating fungus.

6. Nature may reveal its secrets only grudgingly ("God is subtle . . ."), but she obeys orderly natural laws (". . . but He is not malicious.").

TOPIC 2: WHAT ARE PROKARYOTES AND VIRUSES?

KEY POINTS

✓ *What are the main differences between prokaryotes and eukaryotes?*

✓ *What features allow bacteria to adapt to environmental change?*

✓ *What are the relative sizes of eukaryotic cells, prokaryotic cells, and viruses?*

Prokaryotes and **viruses** are too small to see with the naked eye and are the prime focus of the **microbiologist**. Prokaryotes reproduce **asexually**, are **haploid**, and do not have their DNA localized within a nucleus surrounded by a unit membrane. **Eukaryotes** reproduce **sexually** (with the exception of some lower eukaryotes that reproduce sexually and asexually), are not haploid, and possess internal organelles bound by a unit membrane. The genomic DNA of eukaryotic cells is contained in a nucleus surrounded by a membrane. The volume of a bacterium is typically 1000 times less than that of a eukaryote. Larger eukaryotic cells use compartmentalization to localize various activities of the cell.

There are three major phylogenetic **domains** of living organisms based on comparative sequencing of 16S and 18S RNA (see Chapter 7). The domain **Bacteria** is a group of phylogenetically related prokaryotes that are distinct from the group of phylogenetically related prokaryotes in the domain **Archaea**. The Archaea domain contains the methanogens, most extreme halophiles and hyperthermophiles, and the genus *Thermoplasma*. An organism in the Archaea domain is called an **archaeon** (plural, archaea). Eubacteria or bacteria refer to organisms in the

Bacteria domain. All eukaryotes are in the domain **Eukarya**. The microbial eukaryotes (algae, fungi, and protozoa) live mostly as single cells, but some are multicellular (e.g., kelp and mushrooms). Eukaryotes are phylogenetically more closely related to archaea than to bacteria (see Chapter 7).

Viruses are obligate intracellular parasites that can grow and reproduce only when inside their host. Some bacteria are also obligate intracellular parasites, but viruses contain either DNA or RNA (never both), whereas bacteria possess both DNA and RNA. Another distinction between cells and viruses is that cells use DNA to encode their genes, whereas some viruses encode their genes in an RNA molecule. Viruses are small and require an electron microscope for proper visualization. But do not be fooled by small size. A typical virus weighs 0.00000000000001 (10^{-14}) grams and a typical human weighs 72,000 grams, yet some viruses kill humans. To a rough approximation, the density of biologic organisms is $1\,g/cc^3$, so the volume (as well as the weight) of a human is about 10^{19} times the volume of the virus (calculated by dividing 7.2×10^4 by 10^{-14}).

Topic Test 2: What Are Prokaryotes and Viruses?

True/False

1. Because bacteria are single-celled organisms, they do not undergo differentiation in a manner like that of animals.

2. Prokaryotes and higher eukaryotes have the same reproduction process.

Multiple Choice

3. You can typically fit _____ bacteria inside a human cell.
   a. Two
   b. Ten
   c. One hundred
   d. One thousand
   e. One million

4. Bacteria that are obligate intracellular parasites are similar to viruses because they
   a. Are the same size as viruses.
   b. Are both prokaryotes.
   c. Lack cell walls, which viruses also lack
   d. Both require a host to reproduce.
   e. Both consist of DNA and RNA.

5. Archaea
   a. Only reproduce asexually.
   b. Are phylogenetically more closely related to Eukarya than Bacteria are to Eukarya.
   c. Are prokaryotes.
   d. All of the above

Short Answer

6. Biologic organisms are an important part of our ecosystem. Support or refute the concept that size determines the relative harmfulness of one organism to another.

7. Viruses are sometimes considered to be an example of a living system. Explain why you agree or disagree.

Topic Test 2: Answers

1. **True.** Although bacteria do not undergo differentiation in the sense of forming specialized cells as part of a multicellular organism, they can give rise to specialized forms. An example is the formation of an endospore by some bacteria, which is more resistant to environmental challenges (e.g., heat, radiation, drying).

2. **False.** Prokaryotes reproduce only asexually. Higher eukaryotes reproduce sexually. Some lower eukaryotes reproduce sexually, whereas others reproduce asexually. In fact, some lower eukaryotes can reproduce either sexually or asexually (e.g., Baker's yeast).

3. **d.** The volume is about 1000 times greater, thus providing enough volume for 1000 bacteria to fit.

4. **d.** Both require hosts for growth. The other choices do not meet the properties of bacteria and/or viruses.

5. **d.** choices a and c are correct because archaea are prokaryotic, and choice b is correct based on comparisons of 16S and 18S RNA sequences of organisms in the three domains.

6. Clearly, size cannot be used as a predictor of the potential harm one organism may inflict on another. Of course, in the cat-and-mouse game, the mouse would argue about this. Humans can apply their power of reasoning to out-compete much larger animals. And microorganisms invalidate size as a predictor of potential harm in an even more dramatic way.

7. Arguments for agreeing: Viruses form progeny that genetically inherit the properties of the parental viruses. Viruses undergo mutations and evolutionary changes. Argument for disagreeing: Viruses depend on other living organisms for their growth and reproduction and therefore are not living organisms by themselves. Conclusion: This is a philosophical issue best left for others to decide.

TOPIC 3: IMPORTANT EARLY EVENTS IN MICROBIOLOGY

KEY POINTS

✓ *What are the essential requirements for a subject to be considered a science?*

✓ *What problems unique to the study of microorganisms must be solved for microbiology to be a scientific discipline?*

✓ *What were the roles of Antoine Leeuwenhoek, Louis Pasteur, and Robert Koch in establishing microbiology as a science?*

✓ *What are the essential techniques of the microbiologist?*

✓ *How are criteria established for identifying bacteria as causative agents of disease?*

The microscope is an essential tool for visualizing microorganisms because they are too small to visualize with the naked eye. Because microorganisms are very small, special measures must be taken to ensure that there are no unwanted organisms present. Preventing the introduction of microorganisms from the surroundings into an experiment was essential for disproving the spontaneous generation of life theory. If spontaneous generation of life was true, then the microbiologist would have the dilemma of not being able to predict or control his or her experimental work. Another problem faced by microbiologists was isolation of individual organisms in pure cultures for study. Isolating very small organisms presents special challenges. It is easy for a botanist or zoologist to isolate their organisms. Microbiology had three special requirements to be considered as a true science: visualize the life forms being studied, isolate the organisms they were studying into pure cultures, and conduct experiments whose results were not affected by the unpredictable appearance of life forms.

The **simple microscope** (one lens) was developed in the mid-1600s by **Antoine Leeuwenhoek** and allowed visualization of microorganisms. An improvement was made by **Robert Hooke** in the 1700s when he developed the **compound microscope** (two or more lenses) that increased the useful magnification from 250× up to greater than 1000×.

In the 1880s, **Robert Koch** developed a technique for isolating microorganisms and growing them in **pure culture**. A pure culture of microorganisms is a culture in which only one cell type is present. The breakthrough occurred when the wife of a laboratory worker suggested that **agar** be used for constituting solid medium for bacterial growth instead of gelatin (potato slices were used before the introduction of gelatin). Agar was used as a household gelling factor when making jellies. The advantages of agar compared with gelatin are as follows: many fewer bacteria metabolize agar than gelatin and agar remains gelled at a much higher temperature (above 90°C) than gelatin, which liquefies at about 30°C. Isolated colonies that developed on the solid medium surface represented pure cultures of bacteria.

In the 1860s, **Louis Pasteur** performed an experiment that finally **disproved the theory of spontaneous generation**. This is the classic "**swan-neck**" **flask experiment**, in which airborne organisms were prevented from entering the growth medium while allowing unadulterated air access to the bacterial culture. The lack of organism growth could not be attributed to excluding some mysterious vital component that was present in the air and might be the substance that would have otherwise supported the spontaneous generation of life. In fact, Pasteur applied in a very effective way two essential techniques used daily by the microbiologist: **sterilization** and the **aseptic technique**. Sterilization was required to ensure that the growth medium in the swan-neck flask contained no organisms at the beginning of his experiment. He applied the aseptic technique by designing the swan-neck flask so that it prevented the entry of any **unwanted organisms** into his experiment and successfully disproved the theory of spontaneous generation of life.

The "**golden era of microbiology**" was made possible by these techniques. During this period (from the 1880s to approximately the 1920s), numerous organisms were identified to cause diseases. Robert Koch established **four criteria**, called **Koch's postulates**, that provided definitive evidence for an organism to be the **causative agent of a disease**:

1. Always find the microorganism associated with the diseased animal.

2. Isolate the microorganism from the diseased animal and grow in pure culture.

3. Introduce the organism from pure culture into a healthy animal and observe it to become diseased.

4. Reisolate the organism from the diseased animal and confirm that it is the same as the organism previously isolated.

Obligate intracellular parasites can only fulfill Koch's postulates when they can be grown in pure culture. Some organisms are difficult or impossible to grow on artificial medium outside of an animal. *Mycobacterium leprae* (the causative agent of leprosy) can be grown in the armadillo or in the mouse footpad but not on artificial medium. However, there is compelling and irrefutable evidence that it is a disease-causing agent. One should consider Koch's postulates as *sufficient* evidence that an organism causes a disease, but satisfaction of the criteria of Koch's postulates is not *necessary* when identifying a microorganism as the cause of a disease.

Topic Test 3: Important Early Events in Microbiology

True/False

1. Life is considered to have been formed from nonliving substances, in a spontaneous way, thus validating the spontaneous generation of life theory.

2. Fulfillment of Koch's postulates is both necessary and sufficient for an organism to be shown to be the causative agent of a disease.

Multiple Choice

3. Aseptic technique and sterilization were used by _____ in their contributions to the development of microbiology.
   a. Leeuwenhoek
   b. Pasteur
   c. Koch
   d. Two of the above
   e. All three of the above

4. The earliest contribution to the development of microbiology as a science was
   a. The microscope.
   b. Disproving the spontaneous generation of life theory.
   c. Introduction of the aseptic technique.
   d. Sterilization.

Short Answer

5. What was the key feature of the swan-neck experiment performed by Pasteur that provided strong support in disproving the theory of spontaneous generation of life?

6. Were the early contributions of Leeuwenhoek required for Pasteur and Koch to make their later contributions to the development of microbiology as a science? Explain.

Topic Test 3: Answers

1. **False.** Certainly, life originated, during the billions of years of the history of the earth, from nonliving substances. But this was a "once in a lifetime" event, in this case only once during the existence of the world. Life does not spontaneously appear on any regular basis.

2. **False.** They are sufficient but not necessary.

3. **d.** Only Pasteur and Koch required sterilization and aseptic technique for their experiments. Leeuwenhoek only carried out observations on organisms in the environment, which did not require these procedures.

4. **a.** It is not surprising that the observations of microorganisms were reported before experiments were designed to specifically isolate and culture them.

5. The design of the flask was the key factor that provided compelling evidence against the spontaneous generation of life theory.

6. **Yes.** Without evidence for the existence of the microorganisms, there would be no reason to develop experiments for isolating and growing them. Leeuwenhoek's microscope provided the evidence of their existence.

APPLICATION

Advances in microbiology have required sterilization and aseptic technique. These techniques are so commonly used by microbiologists that we sometimes lose sight of what has been made possible by them. To study and characterize bacteria, one must be able to isolate and grow them in culture. Interpreting data obtained with organisms in pure (i.e., only one known microorganism) culture is less difficult than interpreting data from a culture of two or more known organisms.

Pasteur was able to solve the problem of wine spoilage in France by applying a modification of the sterilization technique. He applied isolation techniques to identify the cause of the problem, which was an acetic acid-producing bacterium, which competed with the yeast that was added with the formation of vinegar instead of wine. By mildly heating the liquid that was being processed in the wine making before adding the yeast, enough heat to kill the bacteria but not enough to destroy the desirable properties and flavor of the liquid, he was able to solve the problem. This process is called pasteurization and is applied to other heat-sensitive liquids, like milk. The goal is to greatly reduce the number of harmful bacteria without adulterating the liquid. Harmful bacteria may be either ones that are deleterious to the quality of the product or disease-causing organisms. The goal is *not* sterilization of the liquid, so there may still be living organisms present.

Chapter Test

True/False

1. When a scientific hypothesis is proposed for which there are no experiments to either support or refute the hypothesis, then it is considered as questionable or unacceptable.

2. A pure culture of bacteria is one in which there are one or more known bacteria growing.

3. Sexual reproduction is the only way that eukaryotic cells reproduce, whereas asexual reproduction is the only way that bacteria reproduce.

Multiple Choice

4. Which of the following organisms cannot fulfill Koch's postulates?
   a. Viruses
   b. Bacteria
   c. Protozoa
   d. All of the above

5. Pasteurization is designed to
   a. Kill all living organisms present in the material being treated.
   b. Make material almost sterile.
   c. Reduce the number of potentially harmful organisms.
   d. All of the above

6. Which of the following does *not* apply to viruses?
   a. Are obligate intracellular parasites.
   b. Grow and produce only inside a host cell.
   c. Contain DNA or RNA but never both.
   d. Can never serve as a host for the growth of other organisms or other viruses.
   e. Always use DNA for encoding their genes.

Short Answer

7. List similarities and differences between viruses and bacteria that are obligate intracellular parasites.

8. Why is inductive reasoning used instead of deductive reasoning in the scientific method?

9. Why don't bacteria undergo meiosis?

Essay

10. Why can bacterial cells divide much more rapidly than eukaryotic cells? In fact, *Vibrio natriegens* divides into two cells every 9.8 minutes at 37°C (this is the shortest known doubling time for a bacterium). In your discussion, consider only the affect of the relative sizes of the two cells and assume the cells to be spherical.

Chapter Test Answers

1. **True**

2. **False**

3. **False**

4. **a**

5. **c**

6. **e**

7. Similarity: require a host for growth and reproduction. Differences: size and bacteria have both RNA and DNA, but viruses have one or the other.

8. Scientist take the "show me" approach. Data must validate scientific theories. Otherwise, they are held suspect. Inductive reasoning uses data to develop theories.

9. Meiosis is a part of sexual reproduction. Bacteria reproduce asexually.

10. The maximum growth rate is limited, in part, by the ratio of the surface area of a cell to its volume. The area of the sphere is $4\pi r^2$, the volume is $4/3\pi r^3$, and the surface area to volume ratio is $3/r$. If cell A is one half the size of cell B, then cell A has twice the surface area-to-volume ratio. The larger surface area-to-volume ratio of a smaller cell provides the potential for higher transport rates of nutrients across the membrane for supporting higher growth rates.

Check Your Performance

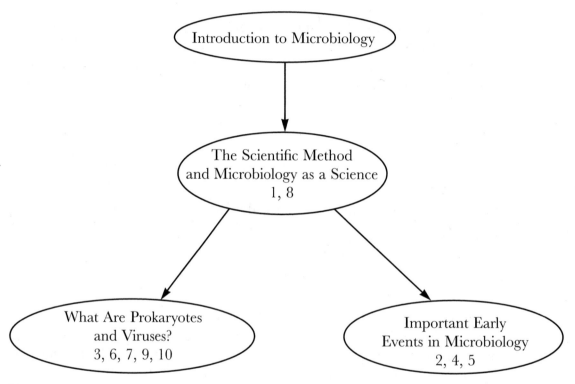

Note the number of questions in each grouping that you got wrong on the chapter test. Identify areas where you need further review and go back to relevant parts of this chapter.

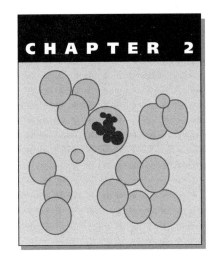

CHAPTER 2

Chemistry Background for Microbiology

Cells can be viewed as chemical factories that use energy from their environment to change molecules from one form to another for synthesis of cellular components and reproduction. Cells must obey the laws of chemistry and physics. In this chapter, chemical principles especially important for understanding the processes used by cells are discussed.

ESSENTIAL BACKGROUND

- Concepts of atoms, ions, molecules, chemical bond, and electronegativity
- Chemical oxidations and reductions
- Meaning of Gibb's free energy
- Biochemistry of macromolecules

TOPIC 1: OXIDATION/REDUCTION REACTIONS

KEY POINTS

✓ *What is electronegativity?*

✓ *What is oxidation? Reduction?*

✓ *Why is a chemical a good oxidant? A good reductant?*

✓ *What are oxidation-reduction (redox) reactions?*

✓ *How do you predict the outcome of the coupling of two half-reactions that form a redox reaction?*

The atom has neutral charge (0) when the number of protons and the number of electrons are equal. When an atom loses one electron it is **oxidized** and has a charge of +1. When an atom gains one electron it is **reduced** and has a charge of −1. If two electrons are lost the ion has charge +2, and so on. **Electronegativity** (introduced by Linus Pauling) is a measure of the electron attracting power of an atom and ranges from 0.8 to 4.0. The electron configurations and electronegativities of the most common atoms in cells are listed in **Table 2.1**. Completed outer electron shells, by the addition or deletion of electrons, result in stable configurations. An **oxidizing agent** reacts by acquiring electrons from the oxidation of another atom or compound (e.g., oxygen acquires two electrons to form a completed outer electron shell). A **reducing agent** causes the reduction of a compound by donating electron(s) to it. Carbon is "on the fence" and can have a completed outer shell by gaining or loosing four electrons.

Table 2.1. Electronic Configurations and Electronegativities of Atoms in Biological Molecules					
		ELECTRON SHELL			
		1	2	3	
ELEMENT	ATOMIC NUMBER	NUMBER OF ELECTRONS IN SHELL			ELECTRONEGATIVITY
Hydrogen (H)	1	1			2.2
Carbon (C)	6	2	4		2.5
Nitrogen (N)	7	2	5		3.1
Oxygen (O)	8	2	6		3.5
Sodium (Na)	11	2	6	1	1
Phosphorous (P)	15	2	8	5	2.1
Sulfur (S)	16	2	8	6	2.4
Chlorine (Cl)	17	2	8	7	2.8

DEFINITIONS

- **Oxidation**: Chemical reaction resulting in the loss of electrons by the reactant.

- **Reduction**: Chemical reaction resulting in the gain of electrons by the reactant.

- **Oxidizer, oxidant, or oxidizing agent**: These terms refer to chemicals that have a high potential for receiving electrons.

- **Reducer, reductant, or reducing agent**: These terms refer to chemicals that have a high potential for loosing electrons.

- **Half-reaction**: A reaction in which the reactant is either oxidized or reduced.
  Example:
  $Na \rightarrow Na^+ + e^-$ (in this half-reaction sodium is oxidized)
  $Cl + e^- \rightarrow Cl^-$ (in this half-reaction chlorine is reduced)

- **Oxidation/reduction (redox) reaction**: A reaction in which electrons are transferred from one chemical to another. Example: $Na + Cl \rightarrow Na^+ + Cl^-$

Because free electrons are extremely reactive, half-reactions always occur as coupled reactions in a redox reaction. One reactant undergoes oxidation and the other undergoes reduction. In general, electronegativity can be used to predict whether a chemical will undergo oxidation or reduction. Atoms with relatively low electronegativities tend to be oxidized and atoms with relatively high electronegativities tend to be reduced.

Topic Test 1: Oxidation/Reduction Reactions

True/False

1. A chemical reaction in which a substance is oxidized will always result in another substance being reduced.

2. Reducer, reductant, and reducing agent are alternative names assigned to chemicals that have a low electronegativity and a high potential for loosing electrons.

Multiple Choice

3. A redox reaction
   a. may result in the making and breaking of covalent bonds.
   b. may give rise to the formation of ionic bonds.
   c. results in the transfer of electrons from one compound to another.
   d. All of the above

4. An atom with high electronegativity tends to
   a. gain electrons.
   b. lose electrons.
   c. form covalent bonds.

Short Answer

5. Briefly explain how the electronegativities of two chemical compounds can be used to predict which will undergo oxidation and which will undergo reduction when the two compounds react with each other. Explain what determines the electronegativity of an atom.

6. Explain why half-reactions always occur in pairs and never alone.

Topic Test 2: Answers

1. **True.** Oxidation results in a loss of electrons. Electrons are highly reactive, and do not remain as free electrons. They will cause the concomitant reduction of another substance.

2. **True.** The three terms are acceptable alternative names assigned to chemicals that readily lose electrons. Chemicals that have low electronegativity more readily lose electrons.

3. **d.** Redox reactions result in electron transfer and formation of covalent or ionic bonds.

4. **a.** Atoms with high electronegativity behave as oxidizing agents and attract electrons.

5. Electronegativity measures the affinity of a substance for electrons. Thus, when two chemicals react, the one with the higher electronegativity will attract electrons from the one with the lower electronegativity.

6. Because no substance can gain electrons without another substance losing electrons.

TOPIC 2: REDUCTION POTENTIAL: A MEASURE OF ENERGY AVAILABLE FROM OXIDATION/ REDUCTION REACTIONS

KEY POINTS

✓ *What is the meaning of the reduction potential of a chemical?*

✓ *How is the reduction potential used to calculate the energy available to bacteria that oxidize chemicals as a source of energy?*

Table 2.2. Reduction Potentials for Some Chemicals

HALF REACTION	E_o (PH 7) (V)
$SO_4^{2-} + 2H^+ + 2e^- \rightarrow SO_3^{2-} + H_2O$	−0.52
$2H^+ + 2e^- \rightarrow H_2$	−0.413
$NAD^+ + H^+ + 2e^- \rightarrow NADH$	−0.32
$S^0 + 2H^+ + 2e^- \rightarrow H_2S$	−0.27
$NO_2^- + 4H^+ + 2e^- \rightarrow NO + H_2O$	0.36
$NO_3^- + 2H^+ + 2e^- \rightarrow NO_2^- + H_2O$	0.43
$2NO_3^- + 2H^+ + 10e^- + 4H_2O \rightarrow N_2 + 10OH^-$	0.74
$2Fe^{3+} + 2e^- \rightarrow 2Fe^{2+}$	0.77
$O_2 + 4H^+ + 4e^- \rightarrow 2H_2O$	0.82

Animal respiration provides energy by a process called **aerobic respiration**. Chemical energy is supplied by redox reactions, in which nutrients are oxidized as a source of fuel and oxygen is reduced, forming water as a waste product. Oxygen is obtained from the environment and serves as the **terminal electron acceptor**. Energy is captured by the vectorial displacement of protons across the mitochondrial membrane. The **proton gradient** formed across the membrane is a source of energy for growth. Bacteria can use a variety of environmental molecules in respiration, including oxygen for aerobic respiration and oxidized forms of carbon (e.g., carbon dioxide), nitrogen (nitrate, nitrite), and sulfur (sulfate, sulfite, etc.) for **anaerobic respiration**.

Redox reactions are a major source of chemical energy during growth by aerobic or anaerobic respiration. The amount of available energy from a redox reaction can be calculated by the change of Gibb's free energy resulting from the redox reaction, which is always negative for spontaneous reactions.

$$\Delta G_o' = -nF\Delta E_o'$$

where n is the number of electrons transferred from the chemical being oxidized, F is Faraday's constant ($96.5\,kJ\,V^{-1}\,mol^{-1}$), and E_o' is the **reduction potential** (note: High values of E_o' indicate a high potential to be reduced, i.e., a high affinity for electrons; **Table 2.2**).

$$\Delta E_o' = (E_o' \text{ of the half-reaction containing the oxidizing agent})$$
$$- (E_o' \text{ of the half-reaction containing the reducing agent})$$

This equation refers to standard-state conditions (25°C and atmospheric pressure) and at pH = 7 (indicated by the "prime"). The E_o of the half-reaction $2H^+ + 2e^- \rightarrow H_2$ at 1 M concentration of H^+ (pH = 0) is arbitrarily set equal to zero. For each increase of one pH unit, E_o becomes more negative by 0.059 V. The E_o' for this half-reaction at pH 7 is calculated by the equation $E_o' = E_o - 0.059 \times pH$, so

$$\Delta E_o' = 0 - 0.059 \times (7) = 0 - 0.413 = -0.413.$$

Topic Test 2: Reduction Potential: A Measure of Energy Available from Oxidation/Reduction Reactions

True/False

1. In a spontaneous chemical reaction, there will always be a decrease in the Gibb's free energy.

2. The reduction potential of the half-reaction donating the electrons in a redox reaction must be less than the reduction potential of the half-reaction that accepts the electrons.

Multiple Choice

3. The change in the Gibb's free energy from a redox reaction depends on
   a. the difference in the reduction potentials of the two half reactions.
   b. the number of electrons transferred between the two half reactions.
   c. All of the above

4. Bacteria obtain energy from respiration by
   a. reduction of external electron acceptors.
   b. oxidation of external electron acceptors.
   c. allowing only one half-reaction to occur.
   d. only under aerobic conditions.

Short Answer

5. Consider bacterium A that oxidizes hydrogen as a source of energy and uses reduction of nitrate to nitrite as the other half-reaction. Now, consider bacterium B that also oxidizes hydrogen as a source of energy but uses reduction of oxygen to water as the other half-reaction (see Table 2.2 for these half-reactions). Which bacterium can obtain more energy from the oxidation of an equivalent amount of hydrogen?

6. Bacteria can obtain energy by the oxidation of chemicals. Using the Gibb's free energy change for an oxidation/reduction (redox) reaction, discuss what variables determine the amount of energy available from a redox reaction.

Topic Test 2: Answers

1. **True.** Changes in Gibb's free energy can be used to predict the outcome of a chemical reaction. If the Gibb's free energy change is positive, then the *reverse* reaction occurs. If the Gibb's free energy change is zero, then the system is at equilibrium, and no changes occur.

2. **True.** The Gibb's free energy change is always negative for a spontaneous chemical reaction.

3. **c.** The equation $\Delta G_o' = - nF\Delta E_o'$ for calculating Gibb's free energy change has two variables: n (the number of electrons transferred in the redox reaction) and $\Delta E_o'$ (the difference in reduction potentials of the two half-reactions).

4. **a.** Respiration requires an available external electron acceptor which is reduced.

5. Bacterium B, because $\Delta E_o'$ is greater.

6. The reduction potentials of the two half-reactions and the number of electrons transferred in the redox reaction.

TOPIC 3: NUCLEIC ACIDS, PROTEINS, AND LIPIDS: STRUCTURES AND ROLES

KEY POINTS

✓ *What are the properties of nucleic acids, proteins, and lipids?*

✓ *What are the roles of nucleic acids, proteins, and lipids?*

✓ *What are the secondary, tertiary, and quaternary structures?*

Nucleic acids are the key for information storage and flow in cell growth. The base sequence of **purines** and **pyrimidines** in DNA codes the genetic information for cells. Ribonucleic acid (RNA) transfers this information during protein synthesis. The purines are adenine and guanine, and the pyrimidines are cytosine and either thymine in DNA or uracil in RNA. These four bases, taken three at a time, code for the amino acid sequence in proteins. The subunits of nucleic acids (deoxyribonucleosides in DNA and ribonucleosides in RNA), are joined by **phosphodiester bonds**. These bonds result in the acidic property and net negative charge at neutral pH of nucleic acids. The concept of pH is reviewed on the Web. There are three levels of structural organization for nucleic acids. Primary structure is the sequence of the bases in the polynucleotide chain. Secondary structure describes its helicity (DNA is a right-handed helix) and whether it is single stranded or double stranded. Single-stranded nucleic acid can have secondary structure by the formation of helical structure between different regions of the polynucleotide chain. Hydrogen bonds contribute to the secondary structure. Tertiary structure applies to covalently closed double-stranded circular nucleic acids. See the Web for more detail.

Cells contain both catalytic proteins (**enzymes**) and structural proteins. Catalytic proteins are called enzymes. The important aspects of catalysts are explained on the Web. Proteins may be acidic, basic, polar, nonpolar, hydrophobic, or hydrophilic depending on their **amino acid** content. Amino acids are joined by a **peptide bond**, which is neither acidic nor basic. Proteins embedded within membranes and exposed at the membrane surface will have both hydrophobic and hydrophilic parts. Molecules that have both hydrophilic and hydrophobic parts are called **amphipathic**. Amphipathic globular proteins typically have an internalized hydrophobic region and a high concentration of hydrophilic amino acids on the surface that is exposed to an aqueous environment. The primary structure of proteins is the amino acid sequence of the polypeptide chain. Secondary, tertiary, and quaternary protein structures result from many weak noncovalent bonds (hydrogen bonds, ionic bonds, etc.) and occasionally covalent disulfide bonds. These determine the final conformation and shape of the protein.

Lipids are a major component of cell membranes. They are amphipathic and form the hydrophobic barrier of the **unit membrane** (see Chapter 3). The hydrophobic part of the lipid consists of **fatty acids** covalently bonded to glycerol molecules by ester bonds. Only water and gases passively diffuse across the membrane.

Topic Test 3: Nucleic Acids, Proteins, and Lipids: Structures and Roles

True/False

1. Proteins contain the information required for the synthesis of nucleic acids.
2. Both lipids and proteins can function as enzymes.

Multiple Choice

3. The phosphodiester bond has _____ charge at neutral pH.
   a. A negative
   b. A neutral
   c. A positive
   d. Either neutral or positive
   e. Either neutral or negative

4. Which of the following (if any) is always hydrophilic?
   a. Nucleic acids
   b. Proteins
   c. Lipids
   d. None of the above

Short Answer

5. Explain why lipids are amphipathic.
6. What causes some proteins to be hydrophobic, others to be hydrophilic, and others to be acidic?

Topic Test 3: Answers

1. **False.** Francis Crick stated the "central dogma" for information flow in biology in the 1960s as DNA ↔ RNA → protein. Protein is not an information source for either DNA or RNA synthesis. Both RNA and DNA can be used as templates to specify the synthesis of new DNA and/or RNA. Viruses provide examples of this.

2. **False.** Well over 99.9% of cellular reactions are catalyzed by proteins (called enzymes). Lipids never serve as catalysts.

3. **a.** The oxygen atoms of the phosphodiester bond have a negative charge.

4. **a.** The chemistry of nucleic acids leads to their hydrophilicity.

5. Lipids have a hydrophobic part, consisting of hydrocarbon chains (fatty acids), and a hydrophilic part, consisting of the glycerol to which the fatty acids are attached. There may be other groups, such as ethanolamine, attached by phosphodiester bonds, which is in the hydrophilic part.

6. The amino acid composition determines this. There are nonpolar, polar, acidic, and basic amino acids.

Chemistry is the study of changes in matter resulting, in part, from the transfer of electrons among chemical reactants. Redox reactions play a major role in these chemical reactions. Bacteria can act as catalysts of these reactions and use the energy derived from the redox reactions for growth. There is a considerable amount of iron associated with mining operations. Some bacteria can oxidize ferrous ion to ferric ion under aerobic conditions. The difference in the reduction potential of the oxidation and reduction half-reactions is only 0.05 V. Therefore, much ferrous ion is oxidized to ferric ion. The ferric ion precipitates, forming ferric hydroxide—$Fe(OH)_3$—called "yellow boy," spontaneously in the aerobic environment with an accumulation of excess protons. The protons accumulate because the hydroxide ions formed from the ionization of water are consumed by the ferric ion precipitation, whereas the protons are not. This results in acidification of the environment and hence acid mine pollution of water bodies into which the drain-off of mine waste dumps occurs.

DEMONSTRATION PROBLEM

Some facultative bacteria (see Chapter 5) can use either nitrate (NO_3^-) or oxygen as a terminal electron acceptor in anaerobic or aerobic respiration. Calculate the available free energy from the oxidation of NADH by nitrate and by oxygen.

Solution

For nitrate: $NADH + NO_3^- + H^+ \rightarrow NAD^+ + NO_2^- + H_2O$

From Table 2.3[1]: $\Delta E_o' = (0.42) - (-0.32) = 0.74\,V$ and $\Delta G_o' = -(2)\,(96.5\,kJ\,V^{-1}\,mol^{-1})\,(0.74\,V) = 142.8\,kJ$

For oxygen: $NADH + \frac{1}{2}O_2 + H^+ \rightarrow NAD^+ + H_2O$

From Table 2.2: $\Delta E_o' = (0.82) - (-0.32) = 1.14\,V$ and $\Delta G_o' = -(2)\,(96.5\,kJ\,V^{-1}\,mol^{-1})\,(1.14\,V) = 220.0\,kJ$

There is a 54% greater energy yield for aerobic respiration.

Chapter Test

True/False

1. Substances always lose electrons when undergoing reduction.

2. A chemical that has a high reduction potential is typically a good oxidizer.

3. Substances with high electronegativities are good oxidants.

4. The available energy from an oxidation/reduction reaction depends on the number of electrons that are transferred between the reactants undergoing oxidation and reduction and also on the difference in the reduction potentials of the reactants.

5. Some nucleic acids are amphipathic.

Multiple Choice

6. Transmembrane proteins span across a membrane. These proteins would be expected to be
   a. acidic proteins.
   b. basic proteins.
   c. amphipathic molecules.

7. When two half-reactions participate in a redox reaction, the half-reaction with the _____ supplies electrons to the redox reaction.
   a. Lower reduction potential
   b. Higher reduction potential
   c. Higher concentration of oxidized chemical
   d. Higher concentration of reduced chemical

8. Amphipathic proteins
   a. can contain only hydrophobic amino acids.
   b. can contain only hydrophilic amino acids.
   c. can contain either only hydrophobic or hydrophilic amino acids.
   d. must contain both hydrophobic and hydrophilic amino acids.

9. When an atom is _____, a _____ ion is formed.
   a. oxidized, negative
   b. oxidized, positive
   c. reduced, negative
   d. reduced, positive
   e. b and c
   f. a and d

10. The reduction potential of the terminal electron acceptor in respiration _____ the reduction potential of the energy source.
    a. must be greater than
    b. must be less than
    c. may be the same as

11. Peptide bonds join
    a. amino acids together to form proteins.
    b. fatty acids to glycerol to form lipids.
    c. deoxyribonucleotides together to form DNA.
    d. None of the above

12. A protein is most suited to be a trasmembrane protein when it is
    a. amphipathic.
    b. hydrophobic.
    c. hydrophilic.

13. Electronegativity can be used to predict the ability of a chemical to
    a. attract electrons.
    b. serve as a reducing agent.
    c. serve as an oxidizing agent.
    d. All of the above

Short Answer

14. Why must an oxidation always be accompanied by a reduction?

15. List the factors that determine the change in Gibb's free energy associated with a redox reaction. Does the concentration of the reactants alter the Gibb's free energy change?

16. Why must the molecules that are contained in membranes be either hydrophobic or amphipathic?

Essay

17. Some bacteria can adapt to changes in their chemical environment by using different external electron acceptors in respiration. Consider a bacterium that has the ability to use oxygen, nitrate, or sulfur as the external electron acceptor, with the formation of water, nitrite, and hydrogen sulfide as waste products, respectively. Explain why this bacterium, when oxidizing hydrogen as a source of energy in the presence of all three of these external electron acceptors, would be expected to consume all the oxygen first, then consume all the nitrate, and finally consume all the sulfur during its growth.

Chapter Test Answers

1. **False**

2. **True**

3. **True**

4. **True**

5. **False**

6. **c**

7. **a**

8. **d**

9. **e**

10. **a**

11. **a**

12. **a**

13. **d**

14. Because free electrons are extremely unstable.

15. The number of electrons transferred in the redox reaction; difference in the reduction potentials of the two half-reactions.

16. The internal portion of the membrane is hydrophobic, and the external surface is hydrophilic. For a molecule to be a component of the membrane, it must either be hydrophobic if internal or amphipathic if a part of the molecule is internal and the other part is exposed at the membrane surface.

17. Use the Gibb's free energy change equation, $\Delta G' = -nF\Delta E_o'$. The largest value for $\Delta E_o'$ occurs when hydrogen and oxygen are the half-reactions, the second largest for hydrogen and nitrate, and the smallest for hydrogen and sulfur. The bacteria regulate their energy metabolism to consume all the electron acceptor that provides the largest ΔG, before beginning to consume the electron acceptor that provides the next largest ΔG. This explains the order in which each is sequentially consumed.

SUGGESTED READINGS

Gerhardt P, Murray RGE, Wood WA, Krieg NR. Methods for general and molecular bacteriology. Washington, D.C.: American Society for Microbiology, 1994.

Segel IH. Biochemical calculations, 2nd edition. New York: John Wiley & Sons, 1976.

Check Your Performance

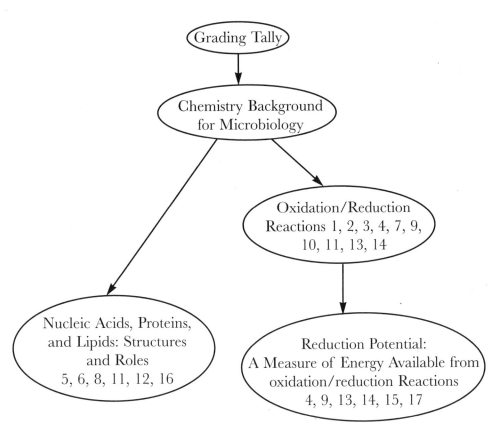

Note the number of questions in each grouping that you got wrong on the chapter test. Identify areas where you need further review and go back to relevant parts of this chapter.

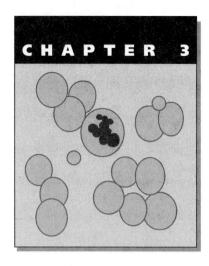

CHAPTER 3

Microbial Structure

$\mathbf{B}$acterial structures are located external and internal to the cyto-plasmic membrane. Some bacterial cytoplasmic structures are surrounded by a so-called membrane (not to be confused with the phospholipid bilayer unit membrane that encloses eukaryotic organelles) containing phospholipid, protein, or glycoprotein.

ESSENTIAL BACKGROUND

- **What is unique about the bacterial cell wall?**
- **What are unit membranes and why are they selectively permeable?**
- **What structures are inside and outside the bacterial cell?**

TOPIC 1: SHAPES AND SURFACE STRUCTURES OF BACTERIA

KEY POINTS

✓ *What are the different bacterial shapes?*

✓ *What are gram-negative and gram-positive bacteria?*

✓ *What is peptidoglycan?*

✓ *What are the properties and functions of bacterial structures?*

The shapes of bacteria are spherical (e.g., **coccus**), rod (e.g., **bacillus**), comma (e.g., **vibrio**), helical (e.g., **spirochete**), filamentous, and appendaged. Bacterial shape is primarily determined by cell wall structure and is one of the properties used to identify and classify bacteria. All bacteria have a cell wall with the exception of the mycoplasma group of bacteria (consisting of six genera), the most common being the genera *Mycoplasma* and *Ureaplasma*. Mycoplasma are short rods whose shape is determined by a robust trilamellar membrane containing sterols.

The **cell wall** of a bacterium consists of **peptidoglycan** layer (also called murein layer), which is a layer that is outside the cytoplasmic membrane. Most bacteria are identified as **gram-positive** or **gram-negative**, depending on their **Gram stain** reaction. The peptidoglycan of gram-negative bacteria is one or two layers thick, and the peptidoglycan of gram-positive bacteria is 20 to 30 layers thick. Gram-negative bacteria have an **outer membrane**; gram-positive bacteria do not. The differences explain the Gram stain reactions of the two bacterial groups.

The peptidoglycan layer is a polymer of alternating *N*-acetylmuramic acid and *N*-acetyl-glucosamine residues, forming long chains that completely surround the cytoplasmic membrane. A tetrapeptide chain is connected to each *N*-acetylmuramic residue. D-Amino acids are found in the tetrapeptide chains. The tetrapeptide chains are joined to one another by a direct transpeptide bond or by a short polypeptide chain of amino acids. The Archaea (see Chapter 7) cell wall consists of **pseudopeptidoglycan**, which is chemically different from peptidoglycan but serves the same functions as the peptidoglycan layer of eubacteria. Some Archaea have an S layer instead of the pseudopeptidoglycan layer.

The outer membrane of gram-negative bacteria consists of two lipid layers and proteins that span across the outer membrane. The inner lipid layer is similar to the lipid layer of the cytoplasmic membrane. The outer lipid layer consists of **lipopolysaccharides** that are polysaccharide chains covalently bound to the lipid molecules and extend outward from the surface. Molecules of less than about 600 to 1000 Da molecular weight can cross the outer membrane by passing through channels in protein molecules called **porins** that span the membrane. The **periplasmic region** is that space bounded by the outer surface of the cytoplasmic membrane and the inner surface of the outer membrane (for gram-negative bacteria) or the cytoplasmic membrane and the peptidoglycan layer in gram-positive bacteria. It is narrow and limited in volume in gram-positive bacteria (less than 10 nm across) and more extensive in gram-negative bacteria (30 to 70 nm). This region contains proteins involved in the interactions of the bacterium with its environment; these proteins are classified into several categories according to their function: solute or ion-binding proteins that function in chemotaxis for sensing and uptake of external nutrients; catabolic enzymes for degrading external polymers into subunit molecules for transport across the cytoplasmic membrane; detoxifying enzymes that protect the bacterium from external toxins; and enzymes for biosynthesis of appendages and structural components external to the outer membrane. Gram-positive bacteria secrete enzymes, called exoenzymes, that usually occur as periplasmic enzymes in gram-negative bacteria.

Features of other bacterial structures external to the cytoplasmic membrane are as follows:

Structure: **Capsules and slime layers**

Molecular composition: polysaccharide or polypeptide

Functions: (1) protection against phagocytosis
(2) retards desiccation (drying out) of bacteria
(3) adherence to surfaces

Other properties: capsules are more compact; slime layers are more diffuse

Structure: **S layer**

Molecular composition: protein and glycoprotein

Functions: surface adhesion; protection against phagocytosis; barrier to extracellular osmotic stress or harmful enzymes.

Other properties: commonly associated with Archae a bacteria and some Bacteria; has a tilelike appearance in electron micrograph. Associated with outer surface of outer membrane of gram-negative bacteria and outer surface of peptidoglycan layer of gram-positive bacteria.

Structure: **Fimbriae** (singular = fimbria)

Molecular composition: protein

Functions: adherence to other cells; adherence to surfaces; attachment to tissue

Other properties: typically there are 100–1000 per bacterium. Diameter = 2–8 nm. Length = 1–5 μm. Fimbriae often consist of a high fraction of nonpolar hydrophobic amino acids. These bacteria will accumulate at an air–oil or air–water interface.

Structure: **Pili** (singular = pilus)

Molecular composition: protein plus small amount of carbohydrate and phosphate.

Functions: Bacterium-to-bacterium attachment during transfer of DNA; attachment to tissue surfaces.

Other properties: One to several per bacterium; length up to 10 μm.

Structure: **Flagella** (singular = flagellum)

Molecular composition: protein

Function: Bacterial motility

Other properties: One to greater than 20 per bacterium. 20 nm diameter and up to 20 μm length. Different arrangements on the bacterium: **monotrichous** = one flagellum at one pole of the bacterium; **amphitrichous** = one flagellum at both poles of the bacterium; **lophotrichous** = tuft of flagella at one or both poles of the bacterium; **peritrichous** = uniform distribution of flagella around the bacterium.

Chemotaxis occurs when bacteria move in response to their chemical. **Positive chemotaxis** describes movement toward a higher concentration of nutrient and **negative chemotaxis** describes bacterial movement away from higher concentrations of toxins. Photosynthetic bacteria use **phototactic** movement to keep them in the region of light required for photosynthesis. Some aquatic bacteria that are microaerophilic (see Chapter 4) carry out **magnetotaxis**, using the Earth's magnetic field for directing movement to remain in a column of water at the location of optimum oxygen concentration.

The flagellum is a relatively rigid helical filament attached to the bacterium by a hook and a basal body consisting of a central rod and several rings that are embedded in the cell surface. The filament is 5 to 20 μm long and is assembled of subunits of one protein with a molecular weight of 50 kDa, called **flagellin**. Elongation of the filament during growth occurs by addition of subunits at the tip. Motility occurs by flagellar rotation using the cytoplasmic proton motive force (see Chapter 5) as the energy source.

In peritrichous and lophotrichous bacteria, the flagellar motor alternates between two states: one generates counterclockwise flagellar rotation associated with bacterial movement along relatively linear trajectories at speed up to 40 μm/s and the other generates clockwise rotation associated with tumbling motility and no net velocity. A signal transduction network regulates the frequency of transition between these two states. Chemotactic bacterial that are in a constant chemical concentration typically move in a random walk of runs of approximately 1 s alternating with tumbles of 0.1 s. When the bacterium senses an increasing concentration of an attractant (positive chemotaxis) or repellant (negative chemotaxis), it tumbles less frequently, thus biasing its random walk in the preferred direction. In monotrichous bacteria, counterclockwise rotation pushes the bacterium in a forward direction that alternates with clockwise rotation that pulls the bacterium in the reverse direction.

Transmembrane **sensory proteins** act as signal transducers by undergoing chemical changes in the region exposed on the cytoplasmic side of the membrane when the region on the external

membrane side interacts with chemical attractants or repellants. Cytoplasmic protein components of the regulatory system are modified in response to the chemical modification of the sensory protein. The modified cytoplasmic protein acts as a regulatory switch for changing the rotational direction of flagella.

Topic Test 1: Shapes and Surface Structures of Bacteria

True/False

1. All gram-positive and all gram-negative bacteria have cell walls.

2. Brownian movement of bacteria is a result of flagella motion.

Multiple Choice

3. _____ are contained in the outer membrane.
   a. Teichoic acids
   b. Lipopolysaccharides
   c. Peptidoglycans
   d. Capsules

4. Which of the following functions to retard the desiccation of a bacterium?
   a. Fimbriae
   b. Capsule
   c. Murein layer
   d. Outer membrane

Short Answer

5. List the benefits of a capsule to a bacterium.

6. Compare and contrast the surface structures of gram-positive and gram-negative bacteria.

Topic Test 1: Answers

1. **True.** The Gram stain response is a direct result of the presence or absence of an outer membrane and of the thickness of the peptidoglycan layer.

2. **False.** Brownian motion is the random nondirectional motion of small particles, such as bacteria.

3. **b.** Lipopolysaccharides are a major component of the outer layer of the outer membrane.

4. **b.** Glycocalyx This refers either to a capsule or slime layer, both of which protect bacteria from desiccation.

5. Retard drying, resist phagocytosis, and facilitate adherence to surfaces.

6. G+: thick peptidoglycan layer (20 to 30 layers); no outer membrane; teichoic acid; limited periplasmic region.
   G−: thin peptidoglycan layer (one or two layers); outer membrane; periplasmic region more extensive than that of G+ bacteria.

TOPIC 2: THE BACTERIAL CYTOPLASMIC MEMBRANE: PROPERTIES AND TRANSPORT SYSTEMS

KEY POINTS

✓ *How do passive diffusion, passive mediated transport, and active transport differ?*

✓ *What is the physical consistency of the membrane when it is at the growth temperature of an organism?*

✓ *What does it mean to be poikilothermic?*

✓ *What happens at the phase transition temperature of a membrane and what properties of the membrane affect its phase transition temperature?*

All cells have a **cytoplasmic membrane** that separates intracellular space from extracellular space because it is a **hydrophobic barrier** for the passage of all polar molecules except water. The membrane is a **lipid bilayer** with the polar parts of the lipid molecules oriented to the outside surfaces and the nonpolar portions in the interior of the membrane. This membrane is referred to as a **unit membrane** that surrounds organelles of eukaryotes and serves as the cytoplasmic membrane of both prokaryotes and eukaryotes.

Prokaryotes are **poikilothermic** organisms whose temperature varies with the environment. Other examples of poikilothermic organisms are single-celled eukaryotic cells, fish, and reptiles. The temperature at which the membrane changes from a gel-like to a fluidlike physical state is called the phase transition temperature. Membranes isolated from bacteria grown at different temperatures are observed to have phase transition temperatures lower than the growth temperature, meaning that membranes are fluidlike during growth (**Table 3.1**). Bacteria adjust the fatty acid composition of their lipids to maintain the membrane in the fluidlike condition at their growth temperature (**Table 3.2**). Longer chain fatty acids and fully saturated fatty acids both contribute to increases in the phase transition temperature because hydrophobic bonding strength between lipid molecules of the membrane is increased by long-chain straight fatty acids.

Substances can cross the membrane by passive diffusion, passive mediated transport, or active transport. Water and gases are the only substances that cross the membrane in significant quantities by passive diffusion. The internal osmotic pressure is adjusted to match the osmotic pressure of the environment by water movement across the membrane. Hypertonic solutions contain a higher salt concentration than the cytoplasm and result in **plasmolysis** during which water moves out of the cell and the cell shrinks. Hypotonic solutions contain a lower salt concentration than the cytoplasm and result in water movement into the cell, cell swelling, and possible burst-

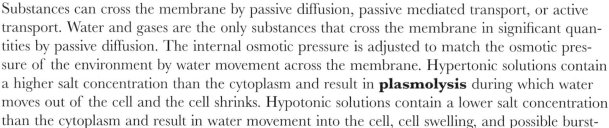

Table 3.1. Phase Transition Temperature of Membranes from *Escherichia coli* Bacteria Grown at Different Temperatures	
GROWTH TEMPERATURE (°C)	PHASE TRANSITION TEMPERATURE (°C)
15	−1
30	16
43	27

Source: Sinesky, Proc Natl Acad Sci 1974;71:522.

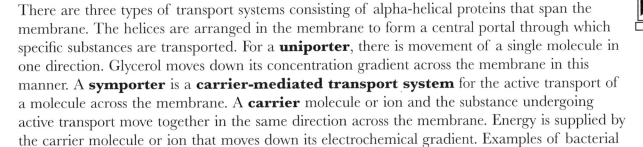

Table 3.2. Fatty Acid Composition of Artificial Membranes with Different Phase Transition Temperatures		
FATTY ACID	CHAIN LENGTH	TRANSITION TEMPERATURE (°C)
Saturated		
Lauric	12	−2
Myristic	14	24
Palmitic	16	42
Stearic	18	58
Mono-unsaturated		
Oleic	18	−22
Di-unsaturated		
Linoleic	18	<−50

ing (lysis). Passive diffusion, active transport, and passive transport are compared in a table on the Web.

There are three types of transport systems consisting of alpha-helical proteins that span the membrane. The helices are arranged in the membrane to form a central portal through which specific substances are transported. For a **uniporter**, there is movement of a single molecule in one direction. Glycerol moves down its concentration gradient across the membrane in this manner. A **symporter** is a **carrier-mediated transport system** for the active transport of a molecule across the membrane. A **carrier** molecule or ion and the substance undergoing active transport move together in the same direction across the membrane. Energy is supplied by the carrier molecule or ion that moves down its electrochemical gradient. Examples of bacterial symporter systems are the phosphate and sulfate symporters that use protons as carrier ions. **Antiporters** act similar to symporters, with the exception that the carrier molecules or ion and the actively transported substance move in opposite directions. Active export of sodium ions occurs by an antiport system that uses protons as carrier ions. **Group translocation** is a transport system that causes chemical change in the substance being transported. The **phosphotransferase system** is a group translocation system for sugars. It consists of a membrane-spanning protein that specifically provides for the transport of the sugar molecule (e.g., glucose, mannitol, or mannose) across the membrane along with phosphorylation of the sugar as it emerges into the cytoplasm by the sequential transfer of a phosphate group from phosphoenolpyruvate to the sugar by a set of three cytoplasmic enzymes. The cytoplasmic enzymes are non-specific for the sugar and the membrane proteins of each **phosphoenolpyruvate:carbohydrate phosphotransferase system** are specific for the sugar being transported.

Topic Test 2: The Bacterial Cytoplasmic Membrane

True/False

1. Membranes with long-chain fatty acids will have a higher phase transition temperature than membranes with short-chain fatty acids.

2. When the salt concentration of the environment of a cell changes, salt moves across the membrane by passive diffusion until the internal salt concentration of the cell is the same as that in the environment.

Multiple Choice

3. Which of the following processes does *not* have saturation of its transport rate?
   a. Active transport
   b. Facilitated diffusion
   c. Passive mediated diffusion
   d. All of the above

4. Some bacteria have optimum growth temperatures around 10°C (called psychrophiles), and others have optimum growth temperatures around 37°C (called mesophiles). To have a fluidlike consistency, the membranes of the psychrophiles will have higher levels of _____ fatty acids than the membranes of the mesophiles.
   a. saturated
   b. unsaturated
   c. both saturated and unsaturated
   d. neither saturated nor unsaturated

Short Answer

5. List two properties of the fatty acids of membrane lipids that alter the phase transition temperature of the membrane and explain why.

6. State two properties that are the same and two properties that are different when comparing passive and active mediated transport systems. List two properties of passive diffusion that distinguish it from *both* passive mediated diffusion and active transport.

Topic Test 2: Answers

1. **True.** There is increased hydrophobic bonding between longer chain fatty acids.

2. **False.** Water is the only molecule that can move across the membrane by passive diffusion that results in change in the internal osmotic pressure.

3. **c.** Passive diffusion does not display saturation kinetics because it does not involve specific sites in the membrane.

4. **b.** Unsaturated bonds in the fatty acid chains result in bends ("kinks") in the aliphatic hydrocarbon chains, reducing the hydrophobic bonding energy between the fatty acids.

5. The length of the fatty acid chains and the degree of saturation of the fatty acid chains both affect the amount of hydrophobic bonding and thus the phase transition temperature.

6. Two properties that are the same: both have saturation kinetics and both transport specific molecules. Two properties that are different: active transport moves molecules against their concentration gradient and passive transport does not and active transport requires energy and passive transport does not. Passive diffusion does not saturate and is nonspecific; neither of these applies to passive mediated and active transport.

TOPIC 3: STRUCTURES FOUND IN THE CYTOPLASM OF BACTERIA

KEY POINTS

✓ *What is a nonunit membrane?*

✓ *What are inclusion bodies?*

✓ *Why do some bacteria have endospores?*

✓ *What structures provide storage of nutrients and energy for bacteria?*

Unlike eukaryotic cells that have unit membrane bound internal organelles, bacteria have **nonunit membrane** bound. Various **inclusion bodies** in the cytoplasm are used for storage of carbon, energy, phosphate, and sulfur. **Gas vacuoles** are clusters of cylindrical **gas vesicles** that are surrounded by a rigid hydrophobic nonunit membrane protein layer that is impermeable to water and only permeable to gas. Their density is much less than the cytoplasm because they are hollow with only gas inside. Gas vacuoles buoy aquatic photosynthetic bacteria toward the top of a water column where there is adequate light intensity for photosynthesis. Aquatic microaerophilic flagellated bacteria contain intracellular magnetic bodies (**magnetosomes**) that cause the cells to align with the Earth's magnetic lines of force to give proper direction to their motion (see previous topic). When growth conditions deteriorate (e.g., when nutrients become depleted or limited), some bacteria form endospores that are resistant to various environmental challenges, such as heat, drying, and radiation. When growth conditions are favorable, endospores germinate into vegetative cells. Endospore formation is not a means of reproduction, because only one spore forms from one vegetative cell (bacteria in the genus *Polyendospora* form multiple endospores, but not as a means of reproduction).

Topic Test 3: Structures Found in the Cytoplasm of Bacteria

True/False

1. All bacterial inclusion bodies are located in the cytoplasm.

2. One function of gas vesicles is to store oxygen for respiration when bacteria are in an anaerobic environment.

Multiple Choice

3. _____ granules are bacterial organelles used for storage of phosphorous.
   a. Glycocalyx
   b. Glycogen
   c. Metachromatic
   d. Polyhydroxybutyrate

4. The membranes of gas vesicles are permeable to
   a. gases.
   b. water.
   c. both gases and water.
   d. neither gases nor water.

Short Answer

5. Outline the endosporulation cycle of bacteria and identify the molecule considered to be essential for the resistance of endospores to environmental challenge and found to be present in endospores in equal proportion to calcium. This molecule is not formed during vegetative growth.

6. What characteristics distinguish unit membranes from nonunit membranes?

Topic Test 3: Answers

1. **True.** Inclusion bodies, by definition, refer to structures internal to the cytoplasmic membrane.

2. **False.** Although gas is present in gas vesicles, the purpose is to prove buoyancy to the bacterium, not as a storage location for gas that is used by the bacterium in growth.

3. **c.** The granules that store phosphate cause a color change in the dye upon staining. These granules are also called phosphate granules.

4. **a.** The membrane of the gas vesicle has proteins on its inner side that are hydrophobic and proteins on its outer face that are hydrophilic.

5. The endospore cycle involves the formation of an endospore in the growing vegetative cell when its environment becomes unfavorable for growth. When conditions are again favorable for growth, the free endospore undergoes germination for the formation of a vegetative cell.

6. Unit membranes consist of a lipid bilayer, exemplified by the cytoplasmic membrane. Organelles in eukaryotes are surrounded by unit membranes. Nonunit membranes consist of protein or other nonlipid material. Organelles in bacteria, if surrounded by a membrane, are surrounded by nonunit membranes.

APPLICATION

This example emphasizes the importance of knowing the structures of a bacterium when designing control methods.

Bacteria are single-celled organisms that are constantly exposed to environmental challenges such as heat, chemicals, and radiation. It is important, when implementing procedures for the control of microorganisms, that endospores are inactivated. In fact, endospores are formed by some bacteria, in response to adverse conditions, and are highly resistant to heat, chemicals, and radiation. Endospores of *Bacillus stearothermophilus* or *Bacillus subtilus* are used as biologic indicators when verifying that a steam autoclave, a gas autoclave, or a dry heat oven is operating properly. When endospores have been killed, one is assured that vegetative bacterial cells are killed.

When using moist heat for killing microorganisms, bacterial endospores were recognized to be a problem in the 1800s by the British physicist John Tyndall. He developed a protocol of boiling a nutrient solution for a few minutes on three or four successive occasions separated by 24 hours at a temperature to encourage germination of endospores into vegetative cells.

The intervening 24-hour intervals permitted the dormant endospores to germinate and become vegetative cells, which were then killed by the 100°C treatment. This is called tyndallization and can be used to sterilize growth medium, when there is concern about exposing the medium to the harsher temperatures of 121°C in the steam autoclave. Endospores are 200-fold more resistant to ionizing radiation, another physical agent used on occasion in the control of bacteria, than the vegetative form.

DEMONSTRATION PROBLEM

Bacteria can be propelled by flagellar action. The velocity of *Bacillus subtilis* is 1.5×10^{-3} cm/s. Now, this is equivalent to 1.5×10^{-5} m/s (or 3.4×10^{-5} miles per hour). Why is this considered a respectively high speed?

Solution

To keep speed in perspective, you should consider the size of the moving object. The dolphin (*Delphinus delphus*) swims at a speed of 1 m/s or at 0.2 body lengths per second. The speed of *B. subtilis* is six body lengths per second. Measured in body lengths per second, the fastest swimmer is the darting fish (*Leuciscus leuciscus*), which has a speed of 175 cm/s and 12 body lengths per second. Protozoa in the genus *Paramecium* move at a speed of 0.1 cm/s, equivalent to five body lengths per second.

Chapter Test

True/False

1. Gram-negative and gram-positive bacteria are distinguished from each other primarily based on differences in the structures of their cytoplasmic membranes.

2. All cells have cytoplasmic membranes.

3. Passive diffusion does not require energy, but passive mediated transport does require energy.

4. The peptidoglycan layer associated with a bacterium is always outside the cytoplasmic membrane of the bacterium.

5. The growth temperature of bacteria is below the temperature at which their cytoplasmic membrane changes from a gel-like to a fluidlike consistency (this is called the phase transition temperature).

6. All bacterial cell walls contain amino acids.

7. Slime layers and capsules have the same functions, even though slime layers are more diffuse than capsules.

8. Capsules, fimbriae, S-layers, pili, and slime layers all contribute to adherence of bacteria to other cells and/or to surfaces.

Multiple Choice

9. Unit membranes are associated with the organelles of
   a. only prokaryotic cells.
   b. only eukaryotic cells.
   c. both eukaryotic and prokaryotic cells.

10. Peptidoglycan contains tetrapeptide chains, which are linked to
    a. *N*-acetylglucosamine residues only.
    b. *N*-acetylmuramic acid residues only.
    c. both *N*-acetylglucosamine and *N*-acetylmuramic acid residues.

11. Osmosis refers to the movement of
    a. ions across the cytoplasmic membrane.
    b. water across the cytoplasmic membrane.
    c. both ions and water across the cytoplasmic membrane.

12. The S-layer of bacteria
    a. contributes to the rigidity of the outer membrane of gram-negative bacteria.
    b. uses polysaccharide for the internal storage of carbon and energy.
    c. is between the cytoplasmic membrane and the peptidoglycan layer.
    d. is part of the bacterial endospore.
    e. contains protein and polysaccharide.

13. The motility of bacteria that are seeking an oxygen concentration that is optimum for their growth is called
    a. aerotaxis.
    b. magnetotaxis.
    c. chemotaxis.
    d. phototaxis.

14. _____ is accompanied by a chemical change in the substance that is transported.
    a. Active transport
    b. Facilitated diffusion
    c. Passive diffusion
    d. Group translocation
    e. Antiport transport

15. Porins are located in the _____ membrane(s) of gram-negative bacteria.
    a. cytoplasmic
    b. inner
    c. outer
    d. inner and outer
    e. inner, outer, and cyplasmic

16. Consider the following bacterial structures and indicate which is(are) located in the cytoplasm.
    (I) capsule   (II) slime layer   (III) granules   (IV) S-layer   (V) flagellum
    a. All of the above
    b. I
    c. II
    d. III

e. IV

f. V

17. Which of the following bacterial structures is associated with the cytoplasmic membrane and in fact is partly imbedded in the cytoplasmic membrane?
(I) capsule (II) slime layer (III) granules (IV) S-layer (V) flagellum

    a. I
    b. II
    c. III
    d. IV
    e. V

Short Answer

18. Discuss why directed bacterial movement operates by alternating between linear "runs" and "tumbles."

19. Among the various structures of a bacterial cell, which one stands out as unique for bacteria and therefore a potential target for the selective control of bacteria by an agent that might be specifically directed against its integrity?

Chapter Test Answers

1. **False**

2. **True**

3. **False**

4. **True**

5. **False**

6. **True**

7. **True**

8. **True**

9. **a**

10. **b**

11. **b**

12. **e**

13. **a**

14. **d**

15. **c**

16. **d**

17. **e**

18. Bacteria undergo directed motion for the purpose of either moving toward desirable environments or away from potentially damaging environments. Because of their very

small size and the manner by which they sense their environment, they must *compare* changes in the environment as they move about. This is achieved by their continual motion, comparing values obtained at succeeding times. When the change occurs in the wrong direction, the bacterium uses the tumbling mode to change its direction, starting off on a new run, and ascertaining that this new direction causes changes in the correct direction. If not, then again, it will use the tumble mode to adjust direction, and so on.

19. Among the structures of the bacterial cell, the cell wall (peptidoglycan layer) is unique to bacteria. This is frequently targeted by antibiotics selective for their control of bacterial growth. In fact, the serendipitous discovery of penicillin as an antibiotic active against bacteria involved its action on the synthesis of bacterial cell walls.

SUGGESTED READINGS

Andrews JH. Comparative ecology of microorganisms. Berlin: Springer-Verlag, 1991.
Neidhart FC, ed. *Escherichia coli* and *Salmonella*, 2nd ed. ASM Press, 1996.
Sinesky. Proc Natl Acad Sci USA 1974;71:522.

Check Your Performance

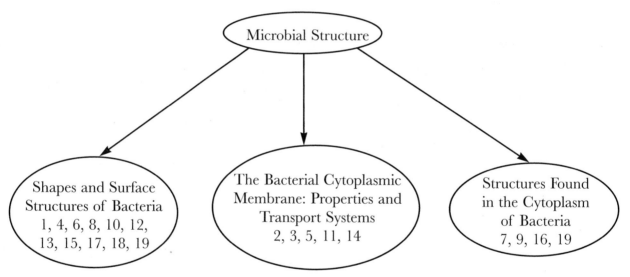

Note the number of questions in each grouping that you got wrong on the chapter test. Identify areas where you need further review and go back to relevant parts of this chapter.

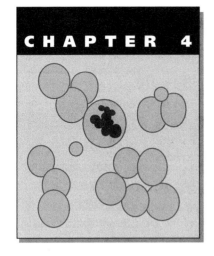

CHAPTER 4

Microbial Growth

Bacteria can grow explosively, accumulating large numbers in a short time. This is why it is important to prevent conditions that would support the growth of harmful organisms in foodstuffs. Toxins from the growth of staphylococci, leading to staphylococcal food poisoning, may accumulate in foods such as ham, chicken salad, and creamed desserts within hours when stored at room temperature. Bacterial reproduction follows a geometric progression, a sequence of numbers (1, 2, 4, 8, 16, . . .) in which the ratio of each number to the immediate preceding number is the same (can you guess what this ratio is?).

ESSENTIAL BACKGROUND

- **Knowledge of exponentials, logarithms, and difference between arithmetic and geometric progressions.**
- **Bacterial colony**
- **Light scattering**

TOPIC 1: QUANTIFYING BACTERIAL GROWTH

KEY POINTS

✓ *What is an arithmetic progression? A geometric progression?*

✓ *What is serial dilution?*

✓ *How do you determine that a bacterium is living?*

✓ *Why do the number of bacteria increase so rapidly during their growth?*

✓ *Why do bacterial suspensions look cloudy?*

Two is the ratio for the geometric progression describing the increase in the number of bacteria during growth and division. The **specific growth rate** for a bacterial culture is the constant used to multiply the previous number of bacteria to determine the next number of bacteria that are present in a culture. For bacteria growing by binary fission in a culture, the number of bacteria present after a given amount of growth time is calculated by the formula $N = N_0 \, 2^n$, where N_0 is the initial number of bacteria and n is the number of divisions during the growth interval.

When measuring the number of bacteria in liquid suspension, the numbers are commonly large. There can be over 1 billion bacteria in each milliliter (mL) of medium. The number 1,200,000,000 can be written in scientific notation as 1.2×10^9, which is convenient when dealing with large numbers. It is assumed that you know how to add, multiply, and divide numbers written in scientific notation. A few examples follow.

$$1.2 \times 10^9 + 1.2 \times 10^9 = 2.4 \times 10^9$$

$$1.2 \times 10^9 + 1.2 \times 10^8 = 1.2 \times 10^9 + 0.12 \times 10^9 = 1.32 \times 10^9$$

$$1.2 \times 10^9 \times 1.2 \times 10^8 = 1.44 \times 10^{17}$$

$$1.2 \times 10^9 / 1.2 \times 10^8 = 1.0 \times 10^1 = 10$$

The number of viable bacteria per mL in a culture is defined as the number of colony forming units per mL. A **colony forming unit** represents a bacterium capable of producing a colony (clump) of cells on the surface of agar medium. Your text and laboratory manual explain that it is desirable to distribute a 0.1- to 1-mL sample volume on an agar surface or into an agar deep such that between about 30 and 300 colonies are formed. Growth of animals and plants can be measured by the growth of size of individuals in the population or by growth in the number of individuals in the population. Bacteria are single-celled organisms. Their growth is measured by the increase in the number of bacteria in a population and not generally by the increase in the size of individual bacteria. Enumerating bacteria requires methods that can detect small cells in the range of $1-5\,\mu m$. Because bacterial growth is measured by the increase in the number of bacteria, the ability of a bacterium to undergo growth and *division* is the working definition of a "living" (viable) bacterium. The most common technique used to detect viable bacteria is the formation of bacterial colonies. A **colony of bacteria** is a localized growth of bacteria on an agar surface, resulting in the formation of a bacterial population (colony) that is sufficiently large to be seen by the unaided eye. Enough time must be allowed for growth and development of the colonies. Two important assumptions are involved when quantifying the number of bacteria by colony formation. First, conditions are assumed to allow each viable bacterium to form a colony. This is called the plating efficiency. Second, each colony is assumed to result from the growth and division of a single bacterium. This condition is met by ensuring a low probability of two bacteria being sufficiently close together on the surface of the agar that would result in the two colonies arising from the bacteria appearing as one.

Other methods for enumerating bacteria include **direct microscopic count**, measuring **bacterial weight**, and measuring the **turbidity** of a liquid suspension of bacteria. Direct microscopic count is tedious and requires a **Petroff-Hausser counting chamber**. Weighing bacteria requires a large number of bacteria for accuracy (the wet weight of a bacterium is about $10^{-12}\,g$). The most widely used and practical technique for measuring the number of bacteria in suspension is measuring turbidity.

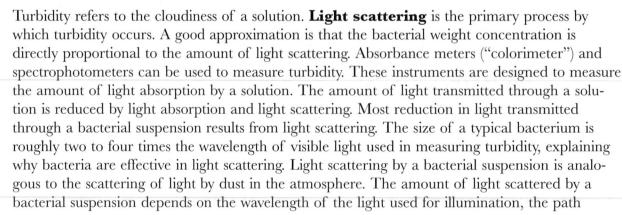

Turbidity refers to the cloudiness of a solution. **Light scattering** is the primary process by which turbidity occurs. A good approximation is that the bacterial weight concentration is directly proportional to the amount of light scattering. Absorbance meters ("colorimeter") and spectrophotometers can be used to measure turbidity. These instruments are designed to measure the amount of light absorption by a solution. The amount of light transmitted through a solution is reduced by light absorption and light scattering. Most reduction in light transmitted through a bacterial suspension results from light scattering. The size of a typical bacterium is roughly two to four times the wavelength of visible light used in measuring turbidity, explaining why bacteria are effective in light scattering. Light scattering by a bacterial suspension is analogous to the scattering of light by dust in the atmosphere. The amount of light scattered by a bacterial suspension depends on the wavelength of the light used for illumination, the path length of the light through the suspension, and the bacterial weight concentration in the suspension. The bacterial weight concentration depends on the number of bacteria per mL and the size of the bacteria.

Approximately the same amount of absorbance (light scattering) is observed for two different bacterial suspensions (suspension A and suspension B), when the bacteria in suspension A are two times larger than the bacteria in suspension B and the concentration of bacteria in suspension A is one half the concentration of bacteria in suspension B. When colony forming units are measured, only viable cells are counted. None of the other methods described above can distinguish between viable and nonviable cells. This should always be kept in mind when interpreting results.

Among the methods discussed, the order of sensitivity from highest to lowest is colony forming units > microscopic count = turbidity > measuring bacterial weight. The volume of liquid in the Petroff-Hausser counting chamber used to enumerate bacteria by direct microscopic count is $0.02\,mm^3$. The volume of $1\,mL$ is $1000\,mm^3$. Reliable data require approximately 100 bacteria be counted. So, for a total of 100 bacteria to be in the field of view in the Petroff-Hausser chamber, there must be 100 bacteria in $0.02\,mm^3$. Now, $0.02\,mm^3 = 0.00002\,mL$. So, there must be 5×10^6 bacteria/mL in the sample. Reliable turbidity data are available only when there is a concentration of 5×10^6 bacteria/mL or greater. When the cell density is greater than approximately 5×10^8 bacteria/mL, dilution is required to obtain reliable turbidity measurements. Colony forming ability is the most sensitive, because applying $0.1\,mL$ to the agar surface of a standard-sized Petri dish requires only 3,000 viable bacteria/mL to yield 300 colonies for reliable statistics.

Topic Test 1: Quantifying Bacterial Growth

True/False

1. A bacterial colony is a localized growth of bacteria.

2. A liquid suspension of bacteria appears cloudy primarily because the bacteria absorb light that passes through the suspension.

Multiple Choice

3. When bacteria are reproducing by binary fission,
   a. there is a constant increase in the number of bacteria during each time interval.
   b. there is a constant fractional increase in the number of bacteria during each time interval.
   c. there is an arithmetic increase in the bacterial population.

4. Which (if any) of the following is most reliable for determining the number of viable bacteria per mL?
   a. Turbidity measurement
   b. Direct microscopic count
   c. Weighing the bacteria
   d. None of the above

Short Answer

5. Discuss why it is unreliable to use turbidity as a measure of the concentration of bacteria in a suspension unless another enumeration method is used to calibrate the turbidity data.

6. State the basic assumptions required when using the number of colonies formed by a bacterial sample to measure the number of viable bacteria present.

Topic Test 1: Answers

1. **True.** Colonies are visible to the unaided eye because of localized cell mass.

2. **False.** Even though turbidity is often measured using an absorbance meter, the turbidity (cloudiness) of the bacterial suspension results from light scattering by the bacteria.

3. **b.** When a number is increasing according to a geometric progression, there is a constant *fractional* increase in the number, not a constant *numerical* increase in the number.

4. **d.** All three methods listed cannot distinguish viable and nonviable bacteria.

5. Turbidity, by itself, cannot report the actual concentration of bacteria in a suspension. Bacterial concentration measurements require another technique, such as direct microscopic count or measuring the colony forming unit/mL.

6. Two important assumptions are fundamental when equating the number of viable bacteria to the number of colonies formed. In this context, a viable bacterium is defined as one that forms a colony of bacteria on the surface of agar. Now, the two assumptions are: conditions are such that every viable bacterium forms a colony and each colony arises from only one bacterium (not two or more which happened to be too close together).

TOPIC 2: ENVIRONMENTAL EFFECTS ON GROWTH

KEY POINTS

✓ *What is water activity? Osmotic pressure? Hydrostatic pressure?*

✓ *Why is oxygen toxic to some organisms?*

✓ *What are the cardinal temperatures of bacterial growth?*

The **cardinal temperatures** of bacteria identify their minimum, optimum, and maximum temperatures of growth. These temperatures depend to some extent on other environmental factors such as pH and available nutrients. The optimum temperature is always closer to the maximum temperature than to the minimum temperature. Bacterial cardinal temperatures vary greatly among genera and species. The values of the minimum, maximum, and optimum growth temperatures of the different categories are summarized in **Table 4.1**. **Psychrotrophs (facultative psychrophiles)** are more widespread than **psychrophiles** because of the broader temperature range in which they grow. Pathogenic organisms are **mesophiles** because their temperature growth range includes the temperature of the human body. The membrane lipids of **thermophiles** are more saturated than those of mesophilic bacteria. Enzymes from **hyperthermophiles** are heat stable. DNA polymerase from a hyperthermophile serves an important role in biotechnology for DNA amplification (see Chapter 20).

Table 4.1. Cardinal Temperatures for Different Categories of Organisms According to Their Growth Temperature Ranges

CATEGORY	MINIMUM GROWTH TEMPERATURE	OPTIMUM GROWTH TEMPERATURE	MAXIMUM GROWTH TEMPERATURE
psychrophile	<0°C	10°–15°C	20°C
psychrotroph (facultative psychrophile)	0°C	15°–20°C	35°C
mesophile	10°C	35°–40°C	45°C
thermophile	40°C	60°–65°C	70°C
hyperthermophile	65°–80°C	85°–105°C	95°–105°C

Bacteria that grow at high hydrostatic pressure but grow optimally at 1 atm are called **barotolerant**. A **barophile** grows optimally at high hydrostatic pressure and may require high pressure for growth. Each species has a pH growth optimum. **Acidophiles** grow optimally between pH 0 and 5.5, **neutrophiles** between 5.5 and 8.0, and **alkalophiles** between 8.0 and 12.0.

Cells are susceptible to changes in the external osmotic pressure. Water moves across the membrane (which is freely permeable to water) to equilibrate the internal osmotic pressure to that of the environment. The **water activity** of a solution is reduced by interaction of water with solute molecules. Water activity is defined as the ratio of the vapor pressure of the solution to the vapor pressure of pure water: $a_w = P_{soln}/P_{water}$. Water activities vary between 0 and 1, and some representative ones are as follows: human blood = 0.995, seawater = 0.98, maple syrup = 0.900, jellies = 0.800, salt lakes = 0.750, cereals, candy, dried fruit = 0.700. Some microorganisms accumulate **compatible solutes** in their cytoplasm as the external osmotic pressure increases (water activity decreases), which reduce the cytoplasmic water activity to that of the environment by increasing the internal solute concentration. Compatible solutes are either synthesized by the organism or obtained from the environment. Examples of compatible solutes are glycerol, sucrose, mannitol, proline, glutamic acid, and other amino acids. **Halotolerant** (**osmotolerant**) bacteria can tolerate some reduction in water activity but grow best in the absence of added solutes. **Halophiles** require reduced water activity for growth. Organisms able to grow in very dry environments due to lack of water (not because of reduced water activity) are called **xerophiles**.

Oxygen is required as the terminal electron acceptor in aerobic respiration (see Chapter 5). Highly reactive toxic intermediates that form during the four-electron reduction of oxygen to water are the **superoxide radical** ($O_2\cdot^-$), **hydrogen peroxide** (H_2O_2), and the **hydroxyl radical** ($OH\cdot^-$). **Catalase** promotes the reaction of two hydrogen peroxides to form water and oxygen. **Superoxide dismutase** catalyzes the reaction of two superoxide radicals to form hydrogen peroxide and oxygen. Working together, these two enzymes detoxify the superoxide radical and hydrogen peroxide to water and oxygen. Peroxidase catalyzes the reduction of hydrogen peroxide to water by transfer of electrons from NADH. **Singlet oxygen** is an energetic state of the normal ground state of oxygen, **triplet oxygen**, which is formed by photooxidation reactions. **Carotenoids** in photosynthetic organisms quench the reaction by absorbing much of the light responsible for photooxidations.

Oxygen is essential for aerobic respiration. Bacteria are diverse and adaptable. The energy yield from the three types of energy metabolism are, in decreasing order, aerobic respiration, anaerobic respiration, and fermentation (see Chapter 5). Classification of bacteria according to oxygen response is as follows:

1. Obligate aerobe: requires oxygen for growth.

2. Microaerophile: grows optimally in reduced oxygen concentration (2–10%). Note: Atmospheric oxygen concentration is 20%.

3. Facultative anaerobe: grows in presence or absence of oxygen. Grows more efficiently in presence of oxygen. (Note: Some texts use facultative aerobe instead of facultative anaerobe.)

4. Aerotolerant anaerobe: does not use oxygen if present. Is not harmed by the presence of oxygen.

5. Obligate anaerobe: does not use oxygen. Killed in the presence of oxygen.

Obligate anaerobes either lack superoxide dismutase or have very low levels. The other classes of bacteria possess both superoxide dismutase and catalase in varying amount, explaining the different oxygen sensitivities of these bacteria. The main role of oxygen is in energy metabolism. It occasionally serves as a source of oxygen atoms in biosynthesis.

Topic Test 2: Environmental Effects on Growth

True/False

1. The optimum temperature is always nearer the maximum temperature than the minimum temperature.

2. The optimal growth pH is a measure of its internal pH.

Multiple Choice

3. Superoxide dismutase and catalase work together to convert the superoxide radical to
   a. hydrogen.
   b. hydrogen peroxide.
   c. oxygen and water.
   d. ozone.

4. When water activity is low, an organism must _____ its _____ solute concentration.
   a. decrease, internal
   b. decrease, external
   c. increase, internal
   d. increase, external

Short Answer

5. a. What distinguishes a barotolerant from a barophilic organism? b. What distinguishes a barophilic from a halophilic organism?

6. List the most likely classifications of a pathogenic organism with regard to its response to environmental temperature, pH, and osmotic pressure.

Topic Test 2: Answers

1. **True.** The profile of cardinal temperatures that plots growth rate versus temperature of growth indicates less difference between the optimum and maximum growth temperatures.

2. **False.** Internal pH remains in the range of 6.0 to 8.0, regardless of external pH.

3. **c**

4. **c.** An organism can only adjust its internal osmotic pressure.

5. **a.** A barotolerant organism grows best at atmospheric pressure but can tolerate and grow at elevated hydrostatic pressure. A barophilic organism grows best at hydrostatic pressures greater than 1 atm and may require elevated pressure for growth. **b.** A barophilic organism grows best at increased *hydrostatic* pressure, and a halophilic organism grows best at increased *osmotic* pressure.

6. Mesophile, neutrophile, osmotolerant.

TOPIC 3: BACTERIAL GROWTH IN BATCH CULTURE AND CONTINUOUS CULTURE

KEY POINTS

✓ *What is required for a bacterium to be considered alive?*

✓ *What is balanced growth?*

✓ *What changes occur in bacterial physiology and metabolism during the growth phase of a batch culture?*

✓ *What is a chemostat? Turbidostat?*

A common method for culturing (growing) bacteria is to add a small number of bacteria ("seed" bacteria) to a large volume of liquid. A reasonable ratio is adding 1 volume of a liquid bacterial suspension with 10^9 bacterial/mL to 100 volumes of sterile growth medium. The freshly transferred bacteria must adapt to the new growth conditions, which may include changes in one or more of the following: temperature, chemical environment, osmotic pressure, pH, and so forth. Growth of bacteria in this manner is referred to as **batch culture**. A batch culture of bacteria undergoes phases of growth (Figure 4.1).

During the **lag phase**, the bacteria are actively metabolizing as they adapt to the changes in their environment. The length of this phase can be reduced by minimizing the changes that occur when the bacteria are transferred from their previous environment to the new environment. For example, if the growth medium into which they are being transferred is preadjusted to the same temperature of the bacteria from their previous environment, then adjustments required for temperature changes are removed. Second, if the bacteria are in the same growth phase as that which immediately follows the lag phase (the exponential phase) at the time they are transferred, they may continue in the exponential growth phase after transfer (assuming another change does not interfere with this). An important point to consider is that failure of the bacteria to increase in viable number cannot be used to conclude that the bacteria are "doing nothing." They are metabolizing to adapt to the new environment, in preparation for entry into the next growth phase.

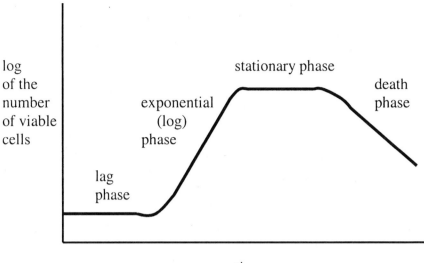

Figure 4.1. Growth phases of a bacterial batch culture.

During **exponential growth phase** (also called log phase or logarithmic phase), the bacteria are actively metabolizing and dividing. Their numbers are increasing at an exponential rate, hence the graph of Figure 4.1 displays a linear increase in the number of viable bacteria in the semilogarithmic plot of viable bacteria/mL against growth time. During this phase there is an explosive increase in the number of viable bacteria. Growth conditions are optimal for reproduction. All components of the bacteria are increasing at the same rate. The rate of increase of each of the components (DNA, RNA, ribosomes, etc.) equals the rate of increase of total biomass and the rate of increase of total cell number. Also, the rate of increases of each component is equal to one another. This is referred to as **balanced growth**. During log phase, the bacteria are physiologically most similar to one another and are carrying out the same metabolic processes.

A bacterial culture in log phase is said to be undergoing **asynchronous growth**, in which each individual bacterium is at its own age, between one division and the next. Some will have just experienced division, producing two "young" bacteria, whereas others are poised to divide and are referred to as "old" bacteria. The concentration of newly divided bacteria is higher than that of bacteria just starting to divide. Researchers frequently do experiments when bacteria are in the log phase of growth because results are most reproducible during this phase when the bacteria are metabolically most similar to one another. Of course, this phase will not last indefinitely as conditions change.

A large number of bacteria accumulate in the batch culture and crowding develops. Waste products accumulate, and one or more essential nutrients may be depleted. These events result in the end of exponential phase and the beginning of **stationary phase**. During stationary phase, there is neither an increase nor a decrease in viable cells. It is incorrect to assume that cell division has stopped. Some bacteria may still be dividing, which is offset by others becoming nonviable, resulting in no change in the viable cell number during stationary phase. As time continues, the conditions that caused the end of log phase and the onset of stationary phase become enhanced, and the final growth phase occurs.

The fourth and last growth phase of a batch culture is **death phase**. There is a decrease in the viable cell number during death phase. Although a few bacteria might still be dividing, many more are losing viability and dying. Toxic waste products accumulate, essential nutrients are depleted, and space for additional formation of new bacteria is limited by overcrowding. The death phase continues until all bacteria have lost their viability or until there remains a subpopulation of bacteria that are resistant to killing under the conditions of the death phase (e.g., endospore formation).

Bacteria grown by **continuous culture** can theoretically be maintained indefinitely in the exponential growth phase. This is achieved by constantly adding fresh medium to the bacterial suspension and removing an equal volume at the same time. By controlling the rate of replacement of bacterial culture with fresh medium (called the dilution rate) and by controlling the concentration of a limiting nutrient being supplied in the fresh medium, it is possible to adjust the growth rate and the cell concentration (density), respectively, of the continuous culture. When the concentration of the limiting nutrient is increased and the dilution rate is held constant, the cell density increases and the growth rate increases. When the concentration of a limiting nutrient is held constant, the cell density remains constant over a fairly broad range of dilution rates, because the growth rate is increasing as the dilution rate increases at a given constant limiting nutrient concentration. A **chemostat** is a continuous culture system in which there is a limiting nutrient and growth rate is determined by the rate at which fresh medium is fed into the culture vessel and an equal volume of effluent is collected. In a **turbidostat**, there is no limiting nutrient, and the flow rate is automatically regulated by a photocell that measures turbidity to maintain a desired cell density. The growth rate remains constant in the turbidostat within a fairly wide range of flow rates. When the flow rate is too rapid, there is "washout" of the cells because the rate of new cell formation cannot keep up with their dilution. At very slow rates, cells can enter the stationary growth phase due to excessive build up of waste products, overcrowding, and/or nutrient depletion. When properly maintained, chemostats and turbidostats provide a steady supply of exponentially growing organisms over a long period of time.

Topic Test 3: Bacterial Growth in Batch Culture and Continuous Culture

True/False

1. During the lag growth phase, cells are neither metabolizing nor dividing.

2. The only growth phase during which bacterial death occurs is the death phase, because this is the only phase during which there is a reduction in the number of viable bacteria.

Multiple Choice

3. In a chemostat,
   a. bacteria are in the exponential growth phase.
   b. there is no limiting nutrient.
   c. the flow rate of fresh medium is controlled by monitoring the cell density.
   d. the growth rate increases as the flow rate of fresh medium decreases.

4. During the exponential growth phase,
   a. bacteria are actively dividing.
   b. a geometric increase in the number of bacteria is taking place.
   c. the highest proportion of bacteria consists of ones which have just been formed by the division of a bacterium.

d. bacteria in the population are physiologically most similar to one another.

e. All of the above

Short Answer

5. Is the lack of increase in the number of viable cells in a batch culture a good indication that the bacteria are *not* metabolizing?

6. Compare and contrast the chemostat and the turbidostat.

Topic Test 3: Answers

1. **False.** Cells are not dividing, but they are metabolizing as they adapt to the new conditions.

2. **False.** Cells may be losing viability during stationary phase.

3. **a.** All other answers do not apply to a chemostat.

4. **e.** During the exponential growth phase, bacteria are dividing, with an increase in viable bacteria. The increase in cell number follows a geometric progression. There are more "young" bacteria than "old" bacteria, and the bacteria are all carrying out the same type of metabolism (i.e., physiologically similar).

5. **No.** During the lag phase of growth, bacteria are not increasing in number but are carrying out metabolism to adjust to their new growth conditions.

6. The chemostat and turbidostat are both continuous culture systems for growing bacteria. The growth rate in the chemostat is controlled by the concentration of a limiting nutrient. There is no limiting nutrient in the turbidostat. The cell density is controlled by a photocell that monitors the turbidity of the culture.

DEMONSTRATION PROBLEMS

1. Antibiotics are widely used. They are usually effective in control of bacterial infections. However, occasionally a mutant bacterium can spontaneously arise that is resistant to the antibiotic. How long would it take for one bacterium that has become resistant to the antibiotic to reach a concentration of approximately 10^6 bacteria/mL blood (a dangerously high level) when its division time is 30 minutes? You know that the total blood volume of an adult male is approximately 13 pints (6 liters).

Solution

The total number of bacteria required to be formed is $6 \times 10^3 \, \text{mL} \times 10^6 \, \text{bacteria/mL} = 6 \times 10^9$ bacteria. Now you must calculate how many divisions are required. Each division doubles the number of bacteria, so this is equivalent to asking how many times you must multiply two by itself to get a number of 6×10^9.

$$2 \times 2 = 2^2 = 4 \quad 2^3 = 8 \quad 2^4 = 16 \quad \ldots \quad 2^{10} = 1024$$

and so on. A quicker way to get the answer is to use the following equation for calculating the number of bacteria after they have divided n times: $N = a \times 2^n$, where N is the number of bacteria after n division and a is the number of initial bacteria. For this example, $N = 6 \times 10^9$, $a = 1$, and n is the unknown.

We must solve the equation $6 \times 10^9 = 2^n$. Use logarithms to solve this equation: $\log(6 \times 10^9) = \log(2^n)$ or $9 \times (\log 6) = n \times (\log 2)$.

$$\log 6 = 0.816 \quad \text{and} \quad \log 2 = 0.301 \therefore 9 \times 0.816 = n \times 0.301 \quad \text{and} \quad n = 24.4$$

In a little over 12 hours, one bacterium dividing every 30 minutes produces a blood concentration of approximately 10^6 bacteria/mL.

2. What is the minimum number of bacteria required for visualization of the colony by the unaided eye? Note: Assume the smallest observable size by the unaided eye is 0.1 mm.

Solution

The number of spherical bacteria closely packed in a single layer that are 0.001 mm (1 μm) diameter to form a circle of 0.1 mm diameter is approximately $(0.1 \, \text{mm})^2/(0.001 \, \text{mm})^2 = 10,000$. But, one layer would not be expected to scatter enough light to be observed. Ten layers would contain approximately 100,000 bacteria, which would provide sufficient light scattering to be visible by the naked eye. Seventeen divisions result in the formation of 130,072 bacteria. Bacteria growing with a 30-minute division time will form this many bacteria in 8.5 hours.

Chapter Test

True/False

1. The chemical composition of the medium changes as growth progresses in a batch culture.

2. Bacteria in batch culture are undergoing balanced growth only during the exponential phase of growth.

3. During the death phase of a batch culture, the reduction in the number of viable cells in the culture is occurring at an exponential rate.

4. The number of bacteria counted in a batch culture using the direct microscopic count method will always be equal to or greater than the number of bacteria counted using the colony forming ability method.

5. Biochemical metabolism is occurring during all four growth phases of a batch culture.

6. Most bacteria reproduce by binary fission, which results in their increase being an arithmetic progression.

Multiple Choice

7. Bacterial division is best described mathematically
   a. by using a linear algebraic equation.
   b. as a constant absolute numeric increase in the number of bacteria.
   c. as a constant fractional increase in the number of bacteria.
   d. as an arithmetic progression.

8. Addition of the logarithms of two numbers is the same as
   a. multiplying the two numbers.
   b. dividing the two numbers.
   c. adding the two numbers.
   d. subtracting the two numbers.

9. Bacteria are most nearly uniform in terms of chemical and physiological properties during the _____ phase.
   a. Lag
   b. Log
   c. Stationary
   d. Death

10. Bacteria growing in batch culture may enter stationary phase as a result of
    a. crowding.
    b. accumulation of toxic waste products.
    c. depletion of an essential nutrient.
    d. Any of these

11. The cardinal temperatures for bacteria are
    a. the minimum temperature at which growth is observed.
    b. the maximum temperature at which growth is observed.
    c. the temperature at which the most rapid rate of growth occurs.
    d. All of the above

12. A microbe growing in the refrigerator that you know is in working order probably is a
    a. mesophile.
    b. psychrophile.
    c. thermophile.
    d. hyperthermophile.

13. Which (if any) of the following measurement techniques will detect bacteria that have undergone lysis?
    a. Direct microscopic count using the Petroff-Hauser counting chamber
    b. Weighing the bacterial mass present
    c. Measuring the absorbance
    d. None of the above

Short Answer

14.* The mass of the earth is estimated to be 5.97×10^{27} g. Consider a bacterium with a mass of 10^{-12} g that is undergoing exponential division, in an unlimited fashion, with a division time of 20 minutes. It has inexhaustible nutrients and unlimited space to grow. How long will it take for enough bacteria to accumulate to equal the mass of the earth, starting with one bacterium?

15. Compare and contrast bacteria that are growing by balanced growth from those that are growing by unbalanced growth.

16. Describe how the length of the lag phase can be shortened.

*Challenge question.

17. During which growth phase would you expect the largest difference in total cell count using viable cell count and direct microscopic cell count methods for enumeration of the bacteria?

Essay

18. Refer to Figure 4.1. The log of the number of viable cells is plotted against growth time. Plot a growth curve for data gathered by turbidity instead of colony forming ability and discuss any differences that you expect between these two growth curves.

Chapter Test Answers

1. **True**

2. **True**

3. **True**

4. **True**

5. **True**

6. **False**

7. **c**

8. **a**

9. **b**

10. **d**

11. **d**

12. **b**

13. **d**

14. This question emphasizes the awesome ability of bacteria to form large numbers of progeny bacteria by exponential growth arising from bacterial reproduction by binary fission. First, divide the mass of the earth by the weight of a single bacterium to determine how many bacteria must be formed:

$$5.97 \times 10^{27} \text{ g}/10^{-12} \text{ g} = 5.97 \times 10^{39} \text{ bacteria}$$

Next, use the equation $N = N_0 2^n$, where n is the number of divisions and N_0 is the number of initial bacteria.

$$N = N_0 2^n, \text{ or } 5.97 \times 10^{39} = 2^n$$

($N_0 = 1$ since we started with one bacterium). Using logarithms:

$$39 \times (\log 5.97) = (n) \times (\log 2);$$
$$39 \times 0.776 = (n) \times 0.301;$$
$$(39 \times 0.776)/0.301 = n;$$

and finally $n = 2010.89$ min, or 33.5 hours! Of course, such growth is not possible.

15. Balanced growth means all essential components, the biomass, and the number of cells are all increasing at the same rate.

16. Shorten the lag phase by minimizing the change in environmental conditions when the bacteria are transferred and by transferring bacteria when they are in the exponential growth phase.

17. The death phase.

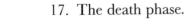

18. The **solid line** is the data from cfu/mL; the **dashed line** is the turbidity data. Differences will appear when there are a significant number of dead cells, which begin to accumulate during the stationary and, especially, the death phases.

SUGGESTED READING

Gerhardt P, Murray RGE, Wood WA, Krieg NR. Methods for general microbiology and molecular bacteriology. ASM Publications, Washington, D.C., 1994.

Check Your Performance

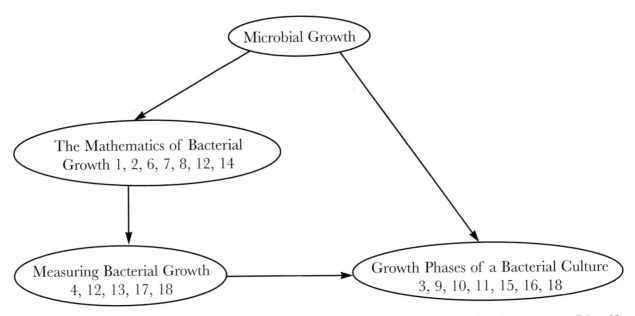

Note the number of questions in each grouping that you got wrong on the chapter test. Identify areas where you need further review and go back to relevant parts of this chapter.

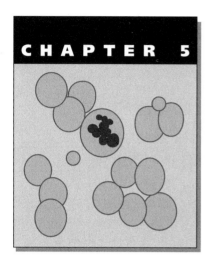

CHAPTER 5

Energy Metabolism: Respiration, Fermentation, and Photosynthesis

Virtually all energy for life on Earth is furnished by the sun, and this energy is captured by photosynthesis that supplies ATP and reducing power for the synthesis of sugars and starch that support the growth of nonphotosynthetic organisms.

ESSENTIAL BACKGROUND

- Oxidation/reduction reactions
- Roles of NAD, NADP, and ATP in energy metabolism
- Properties of respiration and fermentation

TOPIC 1: OVERVIEW OF ENERGY METABOLISM AND CLASSIFICATION OF BACTERIA ACCORDING TO SOURCES OF ENERGY, ELECTRONS, AND CARBON

KEY POINTS

✓ *What categories of compounds are available for bacteria as sources of energy, electrons, and carbon?*

✓ *What is the purpose of electrons obtained from an external source in cell growth?*

✓ *What are the roles of NAD, NADP, and ATP in energy metabolism?*

✓ *What is the meaning of energy charge, anabolic reduction charge, and catabolic reduction charge?*

Energy for bacterial growth is obtained from nutrient chemicals or from light and is conserved for growth in the form of **ATP** and **reducing power**. Metabolism consists of **catabolic** pathways that primarily degrade a substrate and provide energy to generate ATP and reducing power and **anabolic pathways** that use energy for the synthesis of complex molecules from simpler molecules. Pathways that serve the dual functions of anabolism and catabolism are **amphibolic**. Two of the most important amphibolic pathways are the **glycolytic pathway** and the **tricarboxylic acid cycle** (see Chapter 6).

Oxidations that occur during catabolism release electrons that are donated to NAD. Anabolic pathways require a constant supply of both ATP and NADPH as cells synthesize new cellular components and reproduce. There are three important values, calculated as ratios of the alternative forms of ATP, NAD, and NADP, that characterize the state of energy metabolism in a cell and are called, respectively, the **energy charge**, the **catabolic reduction charge**, and the **anabolic reduction charge**:

$$\text{Energy charge} = ([ATP] + 1/2[ADP])/([ATP] + [ADP] + [AMP])$$
$$\text{Catabolic reduction charge} = ([NADH])/([NAD] + [NADH])$$
$$\text{Anabolic reduction charge} = ([NADPH])/([NADP] + [NADPH])$$

All concentrations are expressed as molarities. The factor of 0.5 before ADP accounts for ADP having a high energy bond content one half that of ATP. The oxidized form of NAD is required by catabolic pathways. The reduced form of NADP is required for anabolic pathways. Each ratio has a range of values from 0 to 1. Actively metabolizing cells will maintain the ratios at values favorable for metabolism and growth. Under physiologic conditions, the values of the energy charge hovers around 0.85. The [NADPH]/[NADP] ratio is maintained at a value greater than 0.7 and the [NADH]/[NAD] ratio is maintained at a value less than 0.3. Bacteria use a variety of sources for energy, electrons, and carbon.

A bacterium growing **photolithoautotrophically** uses light as its energy source and inorganic compounds as a source of electrons and carbon. A bacterium using organic compounds as a source of energy and carbon for growth is called a **chemoorganoheterotroph. Autotrophs** obtain their carbon from carbon dioxide and are responsible for synthesizing organic compounds from inorganic carbon dioxide. An external source of electrons is required for chemical reductions that occur during metabolism. A chemical from the environment supplies both energy and electrons for chemotrophs but only electrons for phototrophs. Bacteria that grow as chemolithoheterotrophs are called mixotrophs, obtaining energy from an inorganic compound and carbon from an organic compound. The genus *Beggiotoa* grows mixotrophically by oxidizing reduced inorganic compounds of sulfur for energy and electrons and using an organic compound (acetate) as a source of carbon.

Oxidative phosphorylation (electron transport phosphorylation) refers to the production of ATP from phosphorylation of ADP that is coupled to the transfer of electrons through the **electron transport chain (respiratory chain)**. **Substrate level phosphorylation** refers to phosphorylation of a compound by coupling an **endergonic reaction** to an **exergonic reaction**. Phosphorylation of ADP by transfer of a phosphate group from phosphoenolpyruvate and phosphorylation of glucose to glucose-6-phosphate by transfer of a phosphate group from ATP are examples of substrate level phosphorylation. Phosphorylation of ADP to ATP by energy supplied from light is called **photophosphorylation**.

Topic Test 1: Overview of Energy Metabolism and Classification of Bacteria According to Sources of Energy, Electrons, and Carbon

True/False

1. Light is used as a source of energy and electrons for photosynthesis.

2. Fermentative growth only occurs in the absence of oxygen.

Multiple Choice

3. A bacterium that is using an inorganic compound as its source of electrons is growing
   a. mixotrophically.
   b. chemotrophically.
   c. organotrophically.
   d. lithotrophically.

4. Substrate level phosphorylation
   a. results from the proton motive force (PMF) formed during respiration.
   b. couples an endergonic reaction to an exergonic reaction.
   c. does not occur in photosynthetic organisms.
   d. does not occur while organisms are carrying out respiration.

Short Answer

5. Why will mixotrophs compete best with chemolithoautotrophs and chemoorgano-heterotrophs in an environment where there are both inorganic and organic compounds available as sources of electrons and carbon?

6. Identify the type of growth for plants and animals.

Topic Test 1: Answers

1. **False.** Light only serves as an energy source for photosynthetic organisms. These organisms must obtain electrons from an external chemical (oxygenic photosynthetic organisms use water as their source of electrons).

2. **False.** A common misconception is that fermentation only occurs in the absence of oxygen, perhaps fostered by the observation that the beneficial alcoholic fermentation by yeast only occurs anaerobically. But, this is an adaptive organism. There are many examples of bacteria that lack a respiration chain and grow fermentatively in the presence of oxygen.

3. **d.** The question only identifies the external source of electrons, therefore d is the only acceptable choice.

4. **b.** This choice describes how substrate level phosphorylation occurs. The other choices are incorrect because substrate level phosphorylation can occur simultaneously with oxidative phosphorylation and photophosphorylation.

5. Mixotrophs require both an appropriate inorganic compound (for energy and electron source) and an appropriate organic compound (for carbon source). Chemolithoauto-trophs require only inorganic compounds for energy, electron, and carbon source, and chemoorganoheterotrophs require only organic compounds for energy, electron, and carbon.

6. Plants grow photolithoautotrophically. Mammals grow chemoorganoheterotrophically.

TOPIC 2: ENERGY FROM OXIDATIVE PHOSPHORYLATION AND SUBSTRATE LEVEL PHOSPHORYLATION

KEY POINTS

✓ *What is absolutely required for respiration?*

✓ *What is fermentation?*

✓ *How does respiration create a proton motive force?*

✓ *What is the chemiosmotic theory?*

Respiration refers to the use of an **external electron acceptor** serving as an oxidant. Using an exogenous electron acceptor allows the **net oxidation** of the chemical energy source, providing a much higher energy yield than when there is no net oxidation, as is the case for fermentation (see below). In **aerobic respiration**, oxygen is the external electron acceptor and the waste product is water. In **anaerobic respiration**, something other than oxygen is the external electron acceptor and the waste product depends on what was used as the electron acceptor. The reduction potential for oxygen is the greatest and is the electron acceptor that offers the highest energy yield in respiration (Chapter 2). Other chemicals are used in anaerobic respiration and yield less energy when coupled to a given oxidation reaction compared with oxygen.

The **electron transport chain (respiratory chain)** conserves energy from chemical oxidations by using an **external electron acceptor** and forming a **proton gradient** during the transport of electrons and protons with the ultimate reduction of an **external** electron acceptor. The proton gradient serves as an energy source for ATP synthesis, active transport, and mechanical work. The respiratory chain also maintains the redox concentration balance of NAD+ and NADH.

The **chemiosmotic coupling** of ATP synthesis to the movement of protons across the membrane in response to a proton gradient was a part of the **chemiosmotic theory** introduced by Peter Mitchell in the 1960s to explain the mechanism for using the **PMF** energy formed by the respiratory chain in ATP synthesis. The salient features of the chemiosmotic theory are as follows: the membrane is impermeable to protons, a respiration chain that provides oxidation-reduction-driven proton translocation across the membrane, an external electron acceptor, a proton gradient that forms an electrochemical PMF, and an ATP synthase that uses energy from the PMF.

Fermentation occurs when there is growth without any exogenous electron acceptor. Oxidation of compounds by catabolic pathways reduces NAD+ to NADH, which is oxidized back to NAD+ using an **internal** electron acceptor. Organic compounds serve as both electron donors and acceptors in fermentation. Unifying principles of fermentation include the following: the electron acceptor is internal and not exogenous, only organic compounds are metabolized by fermentation, there is no net oxidation of the growth substrate, the electron acceptor is either pyruvate or a derivative of pyruvate, and ATP is formed only by **substrate level phosphorylation**.

Bacteria carry out different types of fermentation according to the **waste products** produced. Organisms that carry out alcoholic fermentation decarboxylate pyruvate to acetaldehyde and then reduce the acetaldehyde to ethanol. Ethanol and carbon dioxide are the waste products. In lactic acid fermentation, lactic acid is the waste product. Many different waste products are formed by the different fermentation pathways among organisms. Some fermentative pathways provide additional substrate level phosphorylation.

Over a century ago, Louis Pasteur discovered that when yeast cultures metabolizing glucose were exposed to air, the rate of glucose utilization drastically decreased. This phenomenon is called the **Pasteur effect**. Both oxidative phosphorylation and substrate level phosphorylation occur in aerobically grown yeast, whereas only substrate level phosphorylation occurs in yeast growing anaerobically. Because far more energy is obtained from the complete oxidation of glucose during aerobic respiration, oxygen inhibits the phosphofructokinase enzyme of glycolysis (see Chapter 6) and the rate of glycolysis is reduced, which reduces glucose utilization.

Topic Test 2: Energy from Oxidative Phosphorylation and Substrate Level Phosphorylation

True/False

1. Respiration always requires an external electron acceptor.

2. Some bacteria grow by aerobic respiration when oxygen is present or by fermentation when oxygen is absent. The rate of glucose consumption is higher with aerobic respiration because the growth rate is higher.

Multiple Choice

3. During fermentation, ATP is formed from ADP by
   a. substrate level phosphorylation only.
   b. oxidative phosphorylation only.
   c. both oxidative and substrate level phosphorylation.

4. Which of the following is *not* essential for chemiosmotic coupling of the PMF to ATP synthesis?
   a. A membrane impermeable to protons
   b. An external electron acceptor
   c. Substrate level phosphorylation of ADP to ATP
   d. A respiration chain
   e. A proton-driven ATP synthase

Short Answer

5. Why cannot bacteria growing by fermentation use inorganic compounds as an energy source?

6. How do bacteria optimize the aerobic respiration chain for growth in different environments?

Topic Test 2: Answers

1. **True.** The respiratory chain cannot function without using a terminal external electron acceptor.

2. **False.** The Pasteur effect contradicts this statement.

3. **a.** Oxidative respiration requires PMF that is not formed by fermentative reactions.

4. **c.** Substrate level phosphorylation is not a part of the chemiosmotic theory.

5. Because the metabolic intermediates of inorganic compounds are not suitable as endogenous electron acceptors.

6. By using the external electron receptor that has the highest reduction potential.

TOPIC 3: ENERGY FROM LIGHT: PHOTOPHOSPHORYLATION

KEY POINTS

✓ *How does the photosynthetic center convert light energy into chemical energy?*

✓ *What is the role of the membrane in photosynthesis?*

✓ *Does photosynthesis always produce oxygen?*

The sun serves as the primary energy source for all living organisms, and life as we know it would not exist without sunlight. Oxygen, essential for the growth of animals, is provided by photosynthetic organisms. **Light reactions** generate energy by **photosynthesis** and require light. **Dark reactions** occur in the presence or absence of light and fix carbon dioxide into carbohydrates.

The overall oxygenic photosynthetic reaction (light reaction) is:

$$CO_2 + H_2O \rightarrow (CH_2O) + O_2 \text{ (occurs only in the light)}$$

The overall reaction for respiration (dark reaction) is:

$$(CH_2O) + O_2 \rightarrow CO_2 + H_2O \text{ (occurs in the light or dark)}.$$

In order that an organism can grow phototrophically, the rate of the light reaction must be greater than the rate of the dark reaction.

Some highlights that describe photosynthesis in cyanobacteria (blue-green algae), plants, and algae are as follows:

- P680 and P700 are separate photosystems (called photosystem II and photosystem I, respectively), which have photosynthetic reaction centers that maximally absorb photons of 680 and 700 nm wavelength, respectively.

- Water is the electron donor, and oxygen is the waste product.

- Both ATP and reducing energy are formed by the action of oxygenic photosynthesis.

- The components of photosynthesis are membrane associated.

Each reaction center exists in an unexcited and an excited state. In its unexcited state, it behaves as a strong oxidant, receiving electrons from water (P680) or receiving electrons from the electron transport chain (P700). Once having received electrons, the photoreaction center absorbs light and is excited to an energy state with a low reduction potential and now behaves as a strong reductant (substances with low reduction potential are strong reducing agents; Chapter 2).

Photon absorption by the photocenter converts it from a strong oxidizer to a strong reducer and is the manner by which light energy is converted into chemical energy.

There are two modes by which oxygenic photosynthesis operates. In the noncyclic mode, electrons from the oxidation of water are used for the reduction of NADP to NADPH. In the cyclic mode, the electrons received by P700 are cycled back onto P700, with the formation of energy used in the conversion of ADP to ATP during each electron cycle. When there is a surplus of reducing energy (NADPH), the cell can continue to use photosynthesis to obtain energy for ATP formation without further increase in NADPH.

Certain bacteria carry out photosynthesis using an electron donor other than water. This is called anoxygenic photosynthesis. Anoxygenic photosynthesis oxidizes a molecule other than water as the external source of electrons. A molecule other than oxygen is the waste product of this oxidation. When sulfur is the source, sulfate is the waste product. The wavelength of maximum absorption by the photocenter of these bacteria is longer than that of oxygenic photosynthesis and therefore less energy is obtained from the process but is still sufficient to obtain chemical energy for ATP synthesis and reducing power formation. Only one photocenter is contained in anoxygenic photosynthetic bacteria. Anoxygenic photosynthesis is similar to that carried out by photosystem I of oxygenic photosynthesis.

Topic Test 3: Energy from Light: Photophosphorylation

True/False

1. Without photosynthesis, life would soon cease to exist.

2. Dark reactions only occur in the dark.

Multiple Choice

3. Before light is absorbed by the photochemical reaction center of a photosystem, it acts as _____ agent.
   a. an oxidizing
   b. a reducing
   c. both an oxidizing and a reducing agent

4. Oxygenic photosynthesis
   a. forms oxygen as a waste product.
   b. increases the amount of reduced NADP.
   c. uses water as an external source of electrons.
   d. has two photosystems, each of which absorbs photons of different wavelengths.
   e. All of the above

Short Answer

5. Explain why the reduction potential of P700* must be less than the reduction potential of the half reaction $NADP^+ + H^+ + 2\ e^- \rightarrow NADPH$.

6. Why cannot photosystem I of oxygenic photosynthesis oxidize water?

Topic Test 3: Answers

1. **True.** Sunlight is the primary source of energy for the growth of all life forms, and the energy of sunlight is converted into usable chemical energy by photosynthesis. In the food chain, consumers of products of photosynthesis (herbivores for mammals) serve as food supply for the next-in-line consumers (carnivores for mammals). In the world of prokaryotes, chemotrophs consume the chemical energy derived from phototrophs, either directly or by way of biogeochemical cycles (see Chapter 19).

2. **False.** Dark reactions proceed in the presence or absence of light.

3. **a.** The photoreaction center alternates between acting as an oxidizing and a reducing agent. After reduction by the photoreaction center from the oxidation of water, absorption of the photon energy excites the reduced form, providing it a high potential as a reducing agent. Its reduction potential changes from approximately +1.0 to −0.7 V.

4. **e.** All of the choices are properties of oxygenic photosynthesis.

5. For P700* to be energetically able to reduce NADP to NADPH, its reduction potential must have a value less than that of the NADP⁺/NADPH half-reaction. Remember, the lower the value of the reduction potential of a substance, the greater its ability to donate electrons and reduce another substance.

6. The reduction potential is too low (<+0.82 V) to oxidize water. The reduction potential for O_2/H_2O half-reaction is +0.82 V.

APPLICATION

This chapter has reviewed energy metabolism in biologic systems. In fact, life exists because of energy flow. Energy is received as photons of sunlight and reemitted into space from earth as heat. Space serves as the heat sink. Without energy flow, the earth would heat up and life would cease (e.g., greenhouse effect). The energy flow involves the absorption of photons from sunlight, the release of heat, and transfer of this heat to celestial space. Carbon is cycled by assimilation of carbon from carbon dioxide into organic compounds by photosynthetic organisms, using energy from sunlight, and reentry of the carbon as carbon dioxide from metabolism by omnivores, herbivores, carnivores, and microorganisms. Microorganisms result in the ultimate decay of organic and the release of carbon as carbon dioxide.

There is no known naturally occurring organic compound that cannot be metabolized by an organism, under the right conditions, with the release of carbon in the form of carbon dioxide. Carbon dioxide is transferred globally and is the form of carbon that is recycled from dead organisms into new organisms. The fate of all living systems, after death, is that their carbon is available for the birth of new life.

Chapter Test

True/False

1. The concentration of NADP is increased by catabolic pathways.

2. The catabolic reduction charge is increased as the concentration of NADP increases.

3. Noncyclic photosynthesis requires an external source of electrons.

4. A rapid decline in the energy charge of a cell indicates that the cell is dying.

5. In growing cells, the anabolic reduction charge is always greater than the catabolic reduction charge.

6. Mixotrophs are bacteria that are growing as phototrophs.

Multiple Choice

7. For a growing cell, which pair of the following will have values that more closely match one another?
   a. Anabolic reduction charge and catabolic reduction charge
   b. Energy charge and anabolic reduction charge
   c. Energy charge and catabolic reduction charge

8. Consider a cell that is carrying out *only* biosynthetic reactions (of course, such a cell does not exist). For this cell, when carrying out *only* biosynthetic reactions,
   a. the anabolic reduction charge will decrease, and the catabolic reduction charge will remain essentially unchanged.
   b. the anabolic reduction charge will increase, and the catabolic reduction charge will decrease.
   c. both the anabolic and catabolic reduction charges will increase.
   d. both the anabolic and catabolic reduction charges will decrease.

9. When oxygenic photosynthesis is operating in a cyclic mode,
   a. both the anabolic reduction charge and the energy charge are increasing.
   b. only the anabolic reduction charge is increasing.
   c. only the energy charge is increasing.
   d. oxygen accumulates as a waste product.
   e. two of the above are occurring.

10. Organic compounds can be formed from carbon dioxide by
    a. autotrophs only.
    b. heterotrophs only.
    c. organotrophs only.
    d. lithotrophs only.

11. A chemotroph that is growing as a(n) _____ is obtaining its energy from an organic compound.
    a. autotroph
    b. lithotroph
    c. chemotroph
    d. organotroph
    e. heterotroph

12. Fermentation
    a. is used in the production of wine.
    b. occurs only in the absence of oxygen.
    c. is dependent on oxidative phosphorylation.
    d. may be used for the metabolism of either inorganic or organic compounds.

13. Substrate phosphorylation is the exclusive mechanism for ATP synthesis during
    a. aerobic respiration.
    b. anaerobic respiration.
    c. Fermentation.
    d. None of the above

Short Answer

14. Briefly discuss the role of light in the cyclic and noncyclic mode of action of oxygenic photosynthesis.

15. a. How would you classify a bacterium that is using only glucose as its source of energy and carbon for growth?
    b. How would you classify a bacterium that is using only sulfur as its energy source, and using only acetate as its source of carbon? How would you classify this organism if it were using only carbon dioxide as its carbon source?

16. Consider the equation for calculating the energy charge of a cell:

$$\text{Energy charge} = ([ATP] + 1/2[ADP])/([ATP] + [ADP] + [AMP])$$

   Why is the concentration of ADP multiplied by a factor of $1/2$ in the numerator?

17. What is the source of energy for synthesis of ATP by oxidative phosphorylation? Photophosphorylation? Substrate level phosphorylation?

Chapter Test Answers

1. **False**

2. **False**

3. **False**

4. **True**

5. **True**

6. **False**

7. **b**

8. **d**

9. **b**

10. **a**

11. **d**

12. **a**

13. **c**

14. Before light absorption, the photosynthetic center acts as an oxidizing agent. After acquiring electrons and after light absorption, the photosynthetic center acts as a strong reductant. Light provides the energy.

15. **a.** chemoorganoheterotroph. **b.** chemolithoheterotroph (mixotroph); chemolithoautotroph

16. ATP has two phosphodiester bonds, each of which has a large amount of Gibb's free energy available (approximately 33 kJ/mol). ADP has only one phosphodiester bond with this high energy of hydrolysis. Thus, when calculating the energy charge, ADP contributes one half the amount than ATP does to the amount of energy charge.

17. PMF from respiration is the energy source for oxidative phosphorylation. Light is the energy source for photophosphorylation. Coupling an endergonic reaction to an exergonic reaction is the energy source for substrate level phosphorylation.

SUGGESTED READING

Morowitz HJ. Energy flow in biology. Florida: Academic Press, 1968.

Check Your Performance

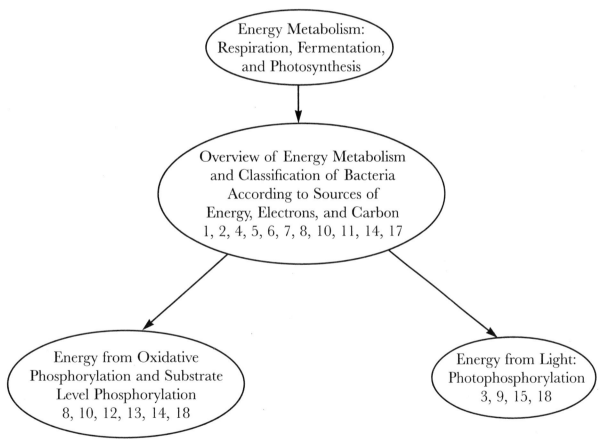

Note the number of questions in each grouping that you got wrong on the chapter test. Indicate areas where you need further review and go back to relevant parts of this chapter.

CHAPTER 6

Catabolic and Anabolic Pathways of Microorganisms

Metabolic processes are categorized as **anabolic** and **catabolic**. Anabolic processes are those that build up components of cell structures. Catabolic processes break down compounds for the release of energy for the cell. Glycolysis, the Entner-Duodoroff pathway, and the tricarboxylic acid cycle break compounds down for energy and supply the precursors for many cellular building blocks. Because they do both anabolic and catabolic functions, these pathways are called **amphibolic**. The processes of these pathways are linked to the electron transport system. See Chapter 5 topic 2 information on how electrons transferred to NAD, NADP, and FAD are processed through the electron transport system.

Metabolic pathways are best shown with drawings and few words.

TOPIC 1: GLYCOLYSIS

KEY POINTS:

✓ *What is the glycolytic pathway?*

✓ *Is oxygen required for glycolysis?*

✓ *How many molecules of ATP are generated by glycolysis?*

Glycolysis is also called the Embden-Meyerhoff pathway or the Embden-Meyerhoff-Parnas pathway for the researchers who worked on it (**Figure 6.1**). **Glycolysis** starts with the glucose molecule. The steps of the pathway as summarized in **Table 6.1** are numbered to match the numbered arrows in Figure 6.1.

Because there are generally two molecules of the 3-carbon compounds, the ATP yield is doubled for a yield of four ATP molecules. There are two molecules of ATP used by the glycolysis process, so the net yield of ATP is two molecules. Glycolysis does not require oxygen.

Topic Test 1: Glycolysis

True/False

1. Glycolysis is an amphibolic pathway.

2. Glycolysis generates precursors for amino acid and lipid synthesis.

3. Phosphenol pyruvate is converted to pyruvic acid.

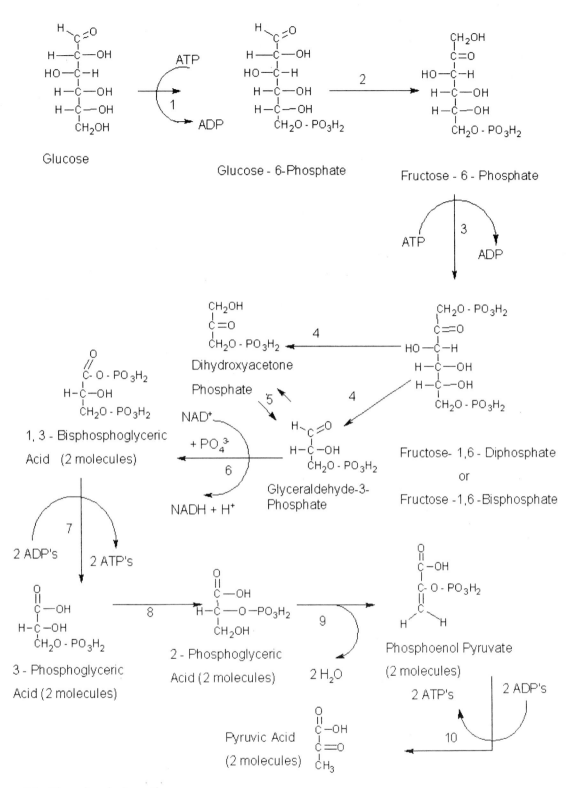

Figure 6.1. The glycolysis pathway.

Table 6.1. Steps of Glycolysis

Step #	Starting Substrate	Enzyme	Changes	Ending Substrate
1	Glucose	Hexokinase or glucokinase	ATP → ADP	Glucose-6-phosphate
2	Glucose-6-phosphate	Phosphoglucose isomerase		Fructose-6-phosphate
3	Fructose-6-phosphate	Phosphofructokinase	ATP → ADP	Fructose-1,6-bisphosphate (aka fructose-1,6-diphosphate)
4	Fructose-1,6-bisphosphate	Aldolase	$NAD^+ \rightarrow$ $NADH + H^+$	Dihydroxyacetone phosphate and glyceraldehyde-3-phosphate (aka triose phosphate)
5	Glyceraldehyde-3-phosphate and dihydroxyacetone phosphate reach equilibrium			
6	Glyceraldehyde-3-phosphate	Glyceraldehyde-3-phosphate dehydrogenase	$NAD^+ \rightarrow$ $NADH + H^+$ $+ PO_4^{3-}$	1,3-Bisphosphoglyceric acid (aka 1,3-diphosphoglyceric acid)

If the dihydroxyacetone phosphate is isomerized to triose phosphate, there will be 2 molecules for the rest of the pathway.

7	1,3-Bisphosphoglyceric acid	Phosphoglycerate kinase	ADP → ATP	3-Phosphoglyceric acid

3-Phosphoglyceric acid can become a precursor molecule for amino acids: serine, cysteine, glycine, and for fatty acids.

8	3-Phosphoglyceric acid	Phosphoglyceric acid mutase		2-Phosphoglyceric acid
9	2-Phosphoglyceric acid	Enolase	$-H_2O$	Phosphoenol pyruvate

Phosphoenol pyruvate can be used as a precursor for amino acids: phenylalanine, tyrosine.

10	Phosphoenol pyruvate	Pyruvate kinase	ADP → ATP	Pyruvic acid

Pyruvic acid is a precursor for the tricarboxylic acid cycle amino acids: alanine, valine, and leucine.

Because there are generally two molecules of the 3-carbon compounds, the ATP yield is doubled for a yield of 4 ATP molecules. There are two molecules of ATP used by the glycolysis process, so the net yield of ATP is two molecules. Glycolysis does not require oxygen.

Multiple Choice

4. Glycolysis uses _____ ATP molecules.
   a. 1
   b. 2
   c. 3
   d. 4

5. Glycolysis generates a net of _____ molecules of ATP.
   a. 1
   b. 2
   c. 3
   d. 4

Short Answer

6. List the molecules that are metabolized in the glycolysis pathway.

Topic Test 1: Answers

1. **True**

2. **True**

3. **True**

4. **b**

5. **b**

6. Glucose, glucose-6-phosphate, fructose-6-phosphate, fructose 1,6-diphosphate, dihydroxyacetone phosphate, glyceraldehyde-3-phosphate, 1,3-bisphosphoglyceric acid (or 1,3-diphosphoglyceric acid), 3-phosphoglyceric acid, 2-phosphoglyceric acid, phosphoenol pyruvate, pyruvic acid.

TOPIC 2: ENTNER-DUODOROFF PATHWAY

KEY POINTS

✓ *What makes the Entner-Duodoroff pathway unique?*

✓ *How does the Entner-Duodoroff pathway differ from glycolysis?*

✓ *How is the Entner-Duodoroff pathway like glycolysis?*

The **Entner-Duodoroff** pathway is unique to prokaryotic cells. This pathway has some steps in common with glycolysis. The Entner-Duodoroff pathway is often considered to be an alternate hexose monophosphate pathway. See web for information on the hexose monophosphate pathway. This pathway starts with glucose as did glycolysis. The numbered steps in **Table 6.2** correspond to the numbered arrows in **Figure 6.2**.

If you have kept track of the ATP molecules used by the pathway, you will notice that the Entner-Duodoroff used only one ATP to phosphorylate the hexose molecules. Because the pathway generates one pyruvic acid molecule directly, there is only one molecule of glyceraldehyde-3-phosphate to continue in the same path as the glyceraldehyde-3-phosphate in glycolysis. Because there is only one triose phosphate to continue the path to pyruvic acid, only one molecule of ATP will be generated at steps 7 and 10. The net is then one ATP molecule, because two were generated by the pathway but one was needed to start the pathway.

Topic Test 2: Entner-Duodoroff Pathway

True/False

1. ATP is generated when 1,3-bisphosphoglyceric acid loses the phosphoryl group from carbon-1.

2. ATP is generated when phosphenol pyruvate loses the phosphoryl group to become pyruvic acid.

3. The Entner-Duodoroff pathway differs from glycolysis when glucose-6-phosphate goes to 6-phosphogluconolactone.

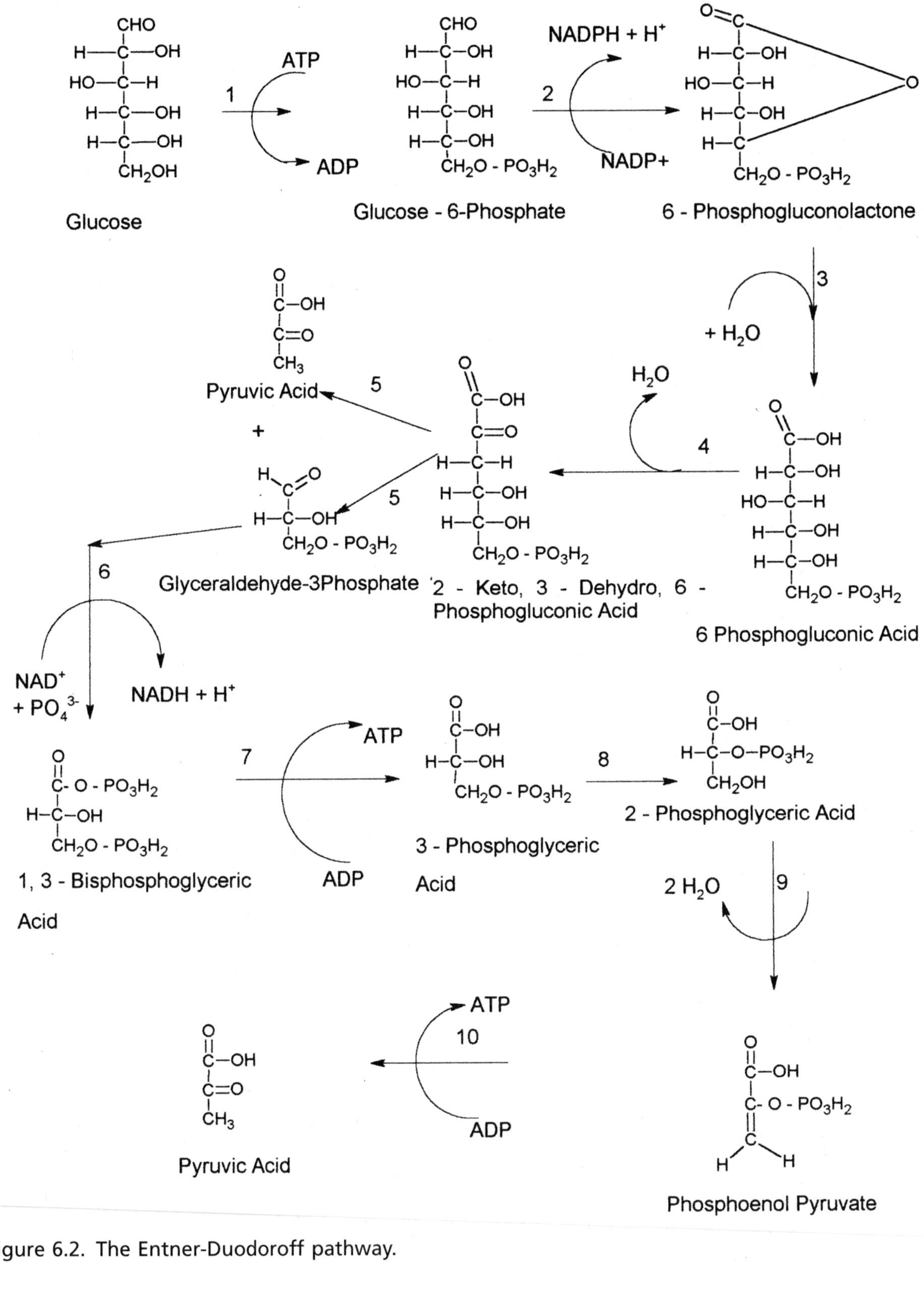

Figure 6.2. The Entner-Duodoroff pathway.

Table 6.2. Steps in Entner-Duodoroff Pathway

STEP #	STARTING SUBSTRATE	ENZYME	CHANGES	ENDING SUBSTRATE
1	Glucose	Hexokinase	$ATP \rightarrow ADP$	Glucose-6-phosphate
2	Glucose-6-phosphate	Glucose-6-phosphate dehydrogenase	$NADP^+ \rightarrow$ $NADPH + H^+$	6-Phosphoglucono-lactone
3	6-Phosphoglucono-lactone	Lactonase	$+H_2O$	6-Phosphogluconic acid
4	6-Phosphogluconic acid	6-Phosphogluconic acid dehydratase	$-H_2O$	2-Keto, 3-dehydro, 6-phosphogluconic acid
5	2-Keto, 3-dehydro,6-phosphogluconic acid	2-Keto, 3-dehydro, 6-phosphogluconic acid aldolase		Glyceraldehyde-3-phosphate and pyruvic acid

The pyruvic acid is a precursor for the tricarboxylic acid cycle or amino acids. The glyceraldehyde-3-phosphate continues in the processing. At this step, the Entner-Duodoroff pathway converges with glycolysis.

STEP #	STARTING SUBSTRATE	ENZYME	CHANGES	ENDING SUBSTRATE
6	Glyceraldehyde-3-phosphate	Glyceraldehyde-3-phosphate dehydrogenase	$NAD^+ \rightarrow$ $NADH + H^+$ $+ PO_4^{3-}$	1,3-Bisphosphoglyceric acid (aka 1,3-diphosphoglyceric acid)
7	1,3-Bisphosphoglyceric acid	Phosphoglycerate kinase	$ADP \rightarrow ATP$	3-Phosphoglyceric acid

3-Phosphoglyceric acid can become a precursor molecule for amino acids: serine, cysteine, glycine, and for fatty acids.

STEP #	STARTING SUBSTRATE	ENZYME	CHANGES	ENDING SUBSTRATE
8	3-Phosphoglyceric acid	Phosphoglyceric acid mutase		2-Phosphoglyceric acid
9	2-Phosphoglyceric acid	Enolase	$-H_2O$	Phosphoenol pyruvate

Phosphoenol pyruvate can be used as a precursor for amino acids: phenylalanine, tyrosine.

STEP #	STARTING SUBSTRATE	ENZYME	CHANGES	ENDING SUBSTRATE
10	Phosphoenol pyruvate	Pyruvate kinase	$ADP \rightarrow ATP$	Pyruvic acid

Pyruvic acid is a precursor for the tricarboxylic acid cycle amino acids: alanine, valine, and leucine.

Multiple Choice

4. The Entner-Duodoroff pathway is considered to be an alternative
   a. tricarboxylic acid cycle.
   b. hexose monophosphate pathway.
   c. glycolysis pathway.

5. The Entner-Duodoroff pathway converges with the _____ pathway for the steps that form pyruvic acid from glyceraldehyde-3-phosphate.
   a. tricarboxylic acid cycle
   b. glycolysis
   c. hexose monophosphate pathway

Short Answer

6. List the names of the molecules metabolized in the Entner-Duodoroff pathway.

Topic Test 2: Answers

1. **True**

2. **True**

3. **True**

4. **b**

5. **b**

6. Glucose, glucose-6-phosphate, 6-phosphogluconolactone, 6-phosphogluconic acid, 2-keto,3-dehydro,6-phosphogluconic acid, glyceraldehyde-3-phosphate, 1,3-bisphosphoglyceric acid (or 1,3-diphosphoglyceric acid), 3-phosphoglyceric acid, 2-phosphoglyceric acid, phosphoenol pyruvate, pyruvic acid.

TOPIC 3: TRICARBOXYLIC ACID CYCLE

KEY POINTS

✓ *Where is most of the energy generated by the tricarboxylic acid cycle?*

✓ *Where is ATP generated directly by the tricarboxylic acid cycle?*

The **tricarboxylic acid cycle** is so named because several of the components of the cycle have three **carboxyl groups** (COOH) on them. Another name for this cycle is the **Krebs cycle** after one of the main researchers who worked on this area of metabolism. It is also called the **citric acid cycle** because the first 6-carbon component of the cycle is citric acid (**Figure 6.3**).

The tricarboxylic acid cycle starts with pyruvic acid from glycolysis or the Entner-Duodoroff pathway. The enzyme that does this is pyruvic acid dehydrogenase. The steps in the tricarboxylic acid cycle are listed in **Table 6.3**.

If you have kept track of the ATP generated by the tricarboxylic acid cycle, you saw that only one ATP is generated directly by the reactions in the cycle. Most of the energy generated by the reactions of the tricarboxylic acid cycle are from the electrons transferred to NAD and FAD. These compounds carry the electrons into the electron transport system where most of the ATP is generated. See Chapter 5 for more information on the electron transport system.

Topic Test 3: Tricarboxylic Acid Cycle

True/False

1. The tricarboxylic acid cycle gets its name from the fact that there are three different carboxylic acids in the cycle.

2. The tricarboxylic acid cycle uses pyruvic acid from the Entner-Duodoroff pathway and glycolysis.

3. *Alpha*-keto-glutaric acid is formed by decarboxylation of isocitric acid.

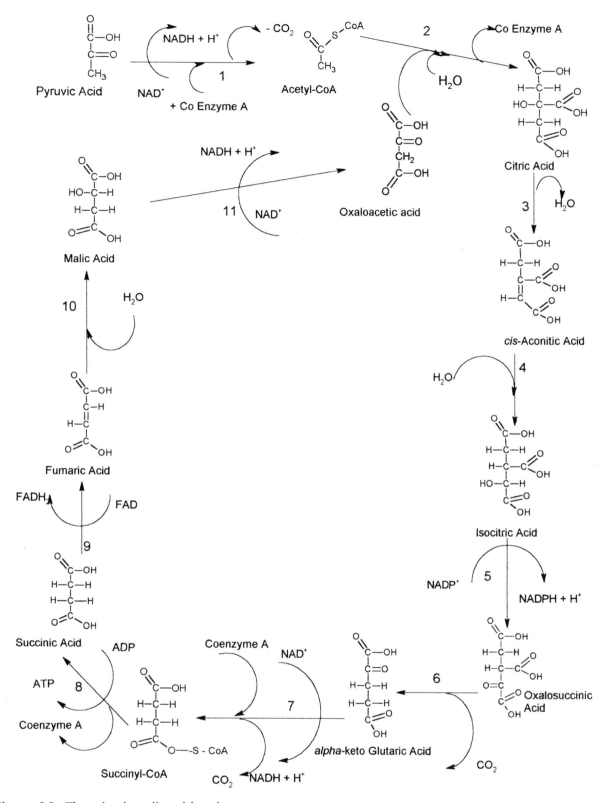

Figure 6.3. The tricarboxylic acid cycle.

Table 6.3. Steps of Tricarboxylic Acid Cycle

STEP #	STARTING SUBSTRATE	ENZYME	CHANGES	ENDING SUBSTRATE
1	Pyruvic acid	Coenzyme A	$NAD^+ \rightarrow$ $NADH + H^+$ $-CO_2$	Acetyl-CoA
2	Acetyl-CoA + H_2O oxaloacetic acid from TCA cycle	Citrate synthetase	Coenzyme A	Citric acid
3	Citric acid loses	Aconitase	$-H_2O$	
4	Cis-aconitric acid	Aconitase	$+H_2O$	Isocitric acid
5	Isocitric acid	Isocitric acid dehydrogenase	$NADP^+ \rightarrow$ $NADPH + H^+$	Oxalosuccinic acid
6	Oxalosuccinic acid	Isocitric acid dehydrogenase	$-CO_2$	Alpha-keto-glutaric acid

Alpha-keto-glutaric acid is a precursor for the amino acids: glutamic acid, glutamine, proline, and arginine.

STEP #	STARTING SUBSTRATE	ENZYME	CHANGES	ENDING SUBSTRATE
7	Alpha-keto glutaric acid	Alpha-keto glutaric acid dehydrogenase complex	$-CO_2$ + coenzyme A $NAD^+ \rightarrow$ $NADH + H^+$	Succinyl-CoA
8	Succinyl-CoA	Succinyl thiokinase	$-$Coenzyme A $ADP \rightarrow ATP$	Succinic acid

Succinic acid is a precursor for heme and porphyrin compounds used in the electron transport system.

STEP #	STARTING SUBSTRATE	ENZYME	CHANGES	ENDING SUBSTRATE
9	Succinic acid	Succinic acid dehydrogenase	$FAD \rightarrow FADH_2$	Fumaric acid
10	Fumaric acid	Fumarase	$+H_2O$	Malic acid
11	Malic acid	Malate dehydrogenase	$NAD^+ \rightarrow$ $NADH + H^+$	Oxaloacetic acid

At this point, the process loops back to enzyme step 2 or the oxaloacetic acid can be used as a precursor for the amino acids: aspartic acid, asparagine, methionine, lysine, threonine, isoleucine.

Multiple Choice

4. How many ATP molecules are generated directly by the tricarboxylic acid cycle?
   a. 0
   b. 1
   c. 2
   d. 3

5. Most of the ATP from the tricarboxylic acid cycle comes from the transfer of electrons to _____ and _____ to carry through the electron transport system.
   a. ATP and ADP
   b. NAD and FAD
   c. FAD and ADP
   d. b and c

Short Answer

6. List the compounds processed in the tricarboxylic acid cycle.

Topic Test 3: Answers

True/False

1. **False.** The tricarboxylic acid cycle gets its name from the fact that there are three carboxyl groups on the citric acid which is the first compound formed.

2. **True**

3. **True**

4. **b**

5. **b**

6. Pyruvate, acetyl-CoA, oxaloacetic acid, citric acid, *cis*-aconitic acid, isocitric acid, oxalosuccinic acid, *alpha*-keto-glutaric acid, succinyl-CoA, succinic acid, fumaric acid, malic acid.

TOPIC 4: CATABOLISM OF PROTEINS AND LIPIDS

KEY POINTS

✓ *How are fatty acids catabolized?*

✓ *How are proteins catabolized?*

Proteins are generally **structural compounds**, so they are not generally used for energy generation. If the cells are in starvation mode, proteins can be broken down into their component amino acids for use. Many of the amino acids need only to have the ammonia group removed (**deaminated**) to become components of one of the energy-generating pathways, such as the tricarboxylic acid cycle or glycolysis. Note the locations in the pathways in topics 1, 2, and 3 or the pentose phosphate pathway table on the web that generate amino acid precursors.

Aromatic amino acids are more difficult for cells to use. These must be broken down into straight chain compounds. Aromatic amino acid degradation steps are shown in **Table 6.4**.

Table 6.4. Steps of Aromatic Amino Acid Catabolism

STEP #	STARTING SUBSTRATE	ENZYME	CHANGES	ENDING SUBSTRATE
1	Phenylalanine	Phenylalanine hydrolase	$NADPH^+ + H^+$ $\rightarrow NADP^+$ $-H_2 + O_2 \rightarrow H_2O$	Tyrosine
2	Tyrosine	Transaminase	$-NH_3$	Parahydroxy phenyl pyruvate
3	Parahydroxy phenyl pyruvate	Parahydroxy phenyl pyruvate oxidase	$-CO + O \rightarrow CO_2$	Homogentisate
4	Homogentisate	Homogentisate oxidase	$+O_2$	4-Maleylacetoacetate
5	4-Maleylacetoacetate	4-Maleylacetoacetate isomerase		4-Fumarylacetoacetate
6	4-Fumarylacetoacetate	4-Fumarylacetoacetate hydrolase	$+H_2O$	Actoacetate and fumaric acid

Fatty acid degradation is a looping process. No matter the length of the fatty acid chain, it receives a coenzyme A molecule and acetyl-CoA is split off from the residue. This process continues until the entire residue is degraded into acetyl-CoA molecules for fatty acids that have an even number of carbons. If the fatty acid has an uneven number of carbon atoms, the final split step will yield acetyl-CoA and propionyl-CoA.

The **fatty acid** degradation process is shown below:

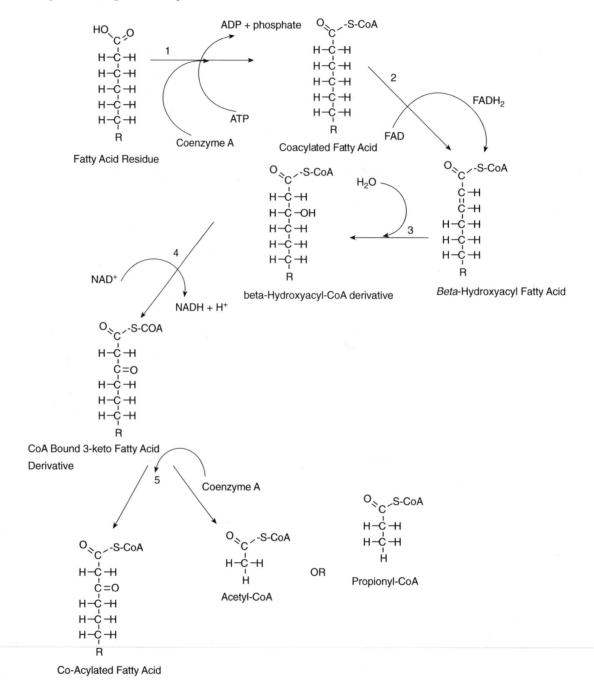

The fatty acid residue can be any length. The letter R in the diagrams represents the part of the chain not shown. The steps of the process are summarized in **Table 6.5**. The process loops back to the beginning and repeats until only acetyl-CoA or propionyl-CoA remains.

Every transfer of electrons to NAD or FAD can go into the electron transport system.

Table 6.5. Steps of Fatty Acid Catabolism				
STEP #	STARTING SUBSTRATE	ENZYME	CHANGES	ENDING SUBSTRATE
1	Residue	Acyl CoA synthetase	+ Coenzyme A ATP → ADP – phosphate	Coacylated fatty acid
2	Coacylated fatty acid	Fatty acid acyl-CoA dehydrogenase	FAD → FADH$_2$	Beta-hydroxyacyl fatty acid
3	Beta-hydroxyacyl fatty acid	3-Hydroxyacyl CoA hydrolase	+H$_2$O	Beta-hydroxyacyl-fatty-acid-acyl-CoA derivative
4	Beta-hydroxyacyl-fatty-acid-acyl-CoA derivative	L-3-Hydroxyacyl CoA dehydrogenase	NAD$^+$ → NADH + H$^+$	CoA bound 3-keto fatty acid derivative
5	CoA bound 3-keto fatty acid derivative	Beta-thiolase	+ Coenzyme A – Acetyl-CoA	Shorter fatty acid residue
The process loops back to the beginning and repeats until only acetyl-CoA or propionyl-CoA remains.				

Start with a fatty acid residue of any length. The letter R in the diagrams represents the part of the chain not shown.

Topic Test 4: Catabolism of Proteins and Lipids

True/False

1. An even number of carbons in a fatty acid chain break down to propionyl-CoA.

2. Coenzyme A is an important part of breaking aromatic amino acids down to straight chain molecules.

3. Proteins are generally structural molecules rather than energy metabolites.

4. Proteins are used for energy generation when the cells are in a starvation state.

Multiple Choice

5. Many amino acids can become components of major energy generating paths such as glycolysis and the tricarboxylic acid cycle by
   a. removal of COOH.
   b. removal of H.
   c. removal of NH$_3$.
   d. removal of O.

6. All fatty acids are broken down by splitting off
   a. acetic acid.
   b. acetyl-CoA.
   c. propionic acid.
   d. oxaloacetic acid.

Topic Test 4: Answers

1. **False.** An even number of carbons in a fatty acid chain break down to acetyl-CoA.

2. **False.** Coenzyme A is not an important part of breaking aromatic amino acids down to straight chain molecules.

3. **True**

4. **True**

5. **c**

6. **b**

APPLICATION

Mapping a metabolic pathway. Ever wonder how we know that carbon-1 is the carbon removed from *alpha*-keto-glutaric acid to form succinic acid in the tricarboxylic acid cycle? The two most likely carbons to be removed are the ones on the ends of the chain, carbon-1 or carbon-5. This was tested by growing cells in culture broth that contained *alpha*-keto-glutaric acid with carbon-1 labeled with C^{14}. After the cells were grown, the carboxylic acids were extracted from the cells. The acids are separated by a method such as thin-layer chromatography and tested to determine whether or not the 4-carbon carboxylic acids were radioactive. Because they were radioactive, researchers knew that carbon-1 was not removed from *alpha*-keto-glutaric acid. The next step is to grow the cells in medium containing 5-C^{14} *alpha*-keto-glutaric acid. After extracting and separating the carboxylic acids, the 4-carbon molecules were found to be nonradioactive, thus confirming carbon-5 was removed.

DEMONSTRATION PROBLEM

Organisms are grown in culture broth containing 1 mole (6.02×10^{23} molecules) of glucose. If all the glucose is used up by the Embden-Meyerhoff pathway, how many moles of ATP will be generated?

Solution

Two moles.

Chapter Test

True/False:

1. In the tricarboxylic acid cycle, phosphoenol pyruvate joins with oxaloacetic acid to form citric acid.

2. *cis*-Aconitic acid is a 5-carbon molecule.

3. Succinic acid is a precursor for heme.

4. Succinic acid is created by decarboxylating *alpha*-keto-glutaric acid.

5. 6-Phosphogluconic acid is a compound created in the glycolysis pathway.

6. 1,3-Bisphosphoglyceric acid is common to both the glycolysis and Entner-Duodoroff pathways.

7. The Entner-Duodoroff pathway creates two molecules of 1,3-bisphosphoglyceric acid from each molecule of glucose.

8. Acetyl-CoA is the final breakdown product of fatty acids that have an even number of carbons.

9. ATP is created from ADP in glycolysis when 1,3-bisphosphoglyceric acid becomes 3-phosphoglyceric acid.

10. Fructose-1,6-bisphosphate breaks down to one molecule each of glyceraldehyde-3-phosphate and pyruvic acid in glycolysis.

11. Glycolysis, tricarboxylic acid cycle, and the Entner-Duodoroff pathways are examples of amphibolic pathways.

12. NAD and FAD can be electron acceptors for both the tricarboxylic acid cycle and glycolysis pathways.

13. Acetyl-CoA is created from pyruvic acid to begin the tricarboxylic acid cycle.

14. 2-Keto, 3-dehydro, 6-phosphogluconic acid is a component of the Entner-Duodoroff pathway.

Multiple Choice

15. Glycolysis requires _____ ATP molecules to phosphorylate the components.
    a. 0
    b. 1
    c. 2

16. ATP is generated by the tricarboxylic acid cycle when electrons are passed to
    a. ADP.
    b. NAD.
    c. FAD.
    d. b and c.

Chapter Test Answers

1. **False**

2. **False**

3. **True**

4. **True**

5. **False**

6. **True**

7. **False**

8. **True**

9. **True**

10. **False**

11. **True**

12. **True**

13. **True**

14. **True**

15. **c**

16. **d**

SUGGESTED READING

Caldwell, Daniel R., Microbial physiology and metabolism, 2nd edition. Belmont, California: Star Publishing Company.

Check Your Performance

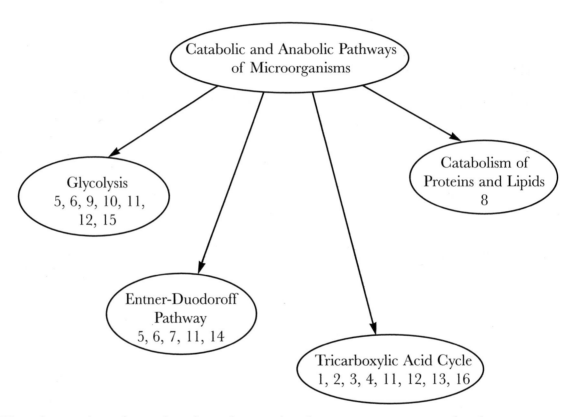

Note the number of questions in each grouping that you got wrong on the chapter test. Identify areas where you need further review and go back to relevant parts of this chapter.

Microbial Taxonomy

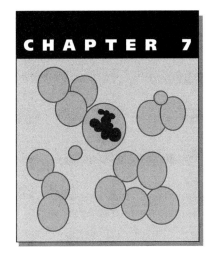

CHAPTER 7

Taxonomy is the science of **classification**. Our current system of nomenclature of plants, animals, protozoa, fungi, and microbes is called **binomial nomenclature**. Binomial means two names. The two names used for microorganisms are the genus and species or specific epithet. We customarily omit the kingdom, phyllum, order, class, and family information when referring to an organism. A common example of the binomial nomenclature is *Escherichia coli*.

The binomial system was created in the 18th century by Carolus Linnaeus, a Swedish botanist. Organisms are classified into groups in the binomial system by their characteristics.

ESSENTIAL BACKGROUND

- **General knowledge of microbiology**

TOPIC 1: PHENETIC AND PHYLOGENIC SYSTEMS

KEY POINTS

✓ *What are some of the characteristics used for phenetic classification?*

✓ *What is phenetic classification?*

✓ *What is phylogenic classification?*

✓ *What is a phenotype?*

Phylogenic systems of classification are based on the **evolution** of the organisms. Because of their small size, fossil evidence of bacteria is hard to come by. Even when fossil evidence is found, bacteria have such simple structures that it is less useful than fossil evidence of more complex organisms. Evolutionary classification of bacteria is also complicated by their rapid rates of reproduction and mutation.

Most classification of bacteria is based on the **phenotype**. Recall that the phenotype is the expression of the organism's **genotype** that can be observed. The Gram stain characteristic is a good example of a phenotype. This classification based on the phenotype/genotype is called **phenetic classification**. The word "phenetic" is derived from combining the words "phenotype" and "genetic."

Organisms are classified phenetically into groups based on phenotypic characteristics of the strains being studied. The optimum number of characteristics to be studied is between 100 and

150, because the number of tests is high enough to use for differentiation without being more than are needed. The tests for the characteristics should cover the entire phenome (phenome is the genotype and phenotype of the organism).

The properties tested in the strains for study can be categorized as follows:

1. Cultural characteristics of the organism;

2. Morphologic characteristics of the organisms;

3. Physiologic characteristics of the organisms, such as pH ranges and growth temperatures;

4. Biochemical characteristics include carbohydrate and nitrogen metabolism;

5. Nutritional requirements look at what the organisms use for their carbon and nitrogen sources;

6. Chemotaxonomic properties deal with the structures of the organisms themselves, such as the chemical composition of the cell walls;

7. Inhibitory tests, such as antibiotic sensitivity;

8. Serologic tests;

9. Genomic characteristics, such as the percentage of guanine (G) and cytosine (C) in the organism's DNA. Mol%GC is the commonly used shorthand for the percentage of G and C. The same mol%GC does not mean the organisms are the same. *Bacillus subtilis* and *Homo sapiens* have the same mol%GC.

Once the tests are done, they are used to classify the organisms. The steps in the phenetic classification are arranging strains to show relationships that are implied by the similarities, estimating the similarities, characterizing the strains, and collecting the strains. Characteristics are equally weighted and relationships by ancestry are not implied in phenetic classification.

Topic Test 1: Phonetic and Phylogenic Systems

True/False

1. Human DNA has the same mol%GC as *Bacillus subtilis* DNA.

2. Phylogenic classification is based on the evolutionary characteristics of the organisms.

3. Phenotype is the genotype's observable expression.

Multiple Choice

4. Biochemical characteristics look at
   a. pH for growth.
   b. NaCl tolerance.
   c. whether the organisms are aerobic or anaerobic.
   d. carbohydrate metabolism.

5. Phenetic classification looks at
   a. genotype of the organisms.
   b. phenotype of the organisms.

c. evolution of the organisms.

d. a and b

Short Answer

6. What is binomial nomenclature?

Topic Test 1: Answers

1. **True**

2. **True**

3. **True**

4. **d**

5. **d**

6. Binomial nomenclature uses two names to identify the organisms. The names used are the genus name and the species name or specific epithet.

TOPIC 2: MATCHING COEFFICIENTS FOR RELATING ORGANISMS AND CONSTRUCTING DENDROGRAMS

KEY POINTS

✓ *What is a dendrogram?*

Dendrograms (from the Greek word for tree), also known as **phenograms** or **dichotamous keys**, can also be used for classification of organisms. Dendrograms are plots of the number of characteristics the strains have in common (**Figure 7.1**).

Many of the tests that would be used for plotting a dendrogram are also commonly found on bacterial identification charts. The charts usually show a characteristic as positive (or negative) if 95% of the strains tested are positive (or negative) for that characteristic.

Common tests for grouping organisms by species are the fermentation of sugars to acid or acid and gas, identification of the carbon and energy sources, utilization of citrate, H_2S production, Gram stain, acid-fast stain, and possibly spore staining characteristics.

In the case of the gram-negative rods, the "IMViC" series—production of *I*ndole, *M*ethyl Red, *V*oges-Proskauer, *C*itrate utilization tests—is useful for differentiating organisms that are alike in most of their other characteristics. For example, *E. coli* is positive for indole and methyl red and negative on the Voges-Proskauer and citrate tests, whereas *Enterobacter aerogenes* is the exact opposite, being negative for indole and methyl red and positive on Voges-Proskauer and citrate.

Additional tests, such as serologic typing, biotyping, antimicrobial sensitivity, phage typing, and bacteriocin typing, are used to group organisms at the subspecies level. This is important in epidemiology to determine what subtype of the organism is responsible for the outbreak of a

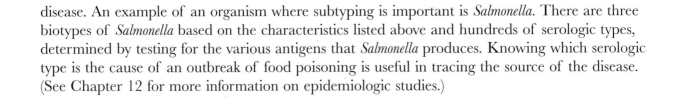

Figure 7.1. The dendrogram above shows the comparison of 10 strains of organisms. Strains A, C, G, and I have a 95% match with each other in the characteristics tested. Strains B, E, and H have a 90% match with each other in the characteristics tested. Strains D, F, and J have an 85% match with each other in the characteristics tested. Strains A, C, G, and I have a 75% match with strains B, E, and H in the characteristics tested. All the strains have a 60% match with each other in the characteristics tested.

disease. An example of an organism where subtyping is important is *Salmonella*. There are three biotypes of *Salmonella* based on the characteristics listed above and hundreds of serologic types, determined by testing for the various antigens that *Salmonella* produces. Knowing which serologic type is the cause of an outbreak of food poisoning is useful in tracing the source of the disease. (See Chapter 12 for more information on epidemiologic studies.)

Topic Test 2: Matching Coefficients for Relating Organisms and Constructing Dendrograms

True/False

1. Dendrograms are used to show the groupings of organisms.

2. The "IMViC" series is used to differentiate gram-positive organisms.

3. *E. coli* is negative for indole and methyl red and positive for Voges-Proskauer and citrate.

4. *Salmonella* has hundreds of different biotypes.

5. Subtyping is important in epidemiologic studies.

Topic Test 2: Answers

1. **True**

2. **False.** The "IMViC" series is used to differentiate gram-negative organisms.

3. **False.** *E. coli* is positive for indole and methyl red and negative for Voges-Proskauer and citrate.

4. **False.** *Salmonella* has hundreds of different serologic types.

5. **True**

TOPIC 3: USE OF rRNA GENES IN PHYLOGENIC SYSTEMS

KEY POINTS

✓ *What are the components of rRNA?*

✓ *Which subunits are used for phylogenetic studies?*

Genetic ancestry of bacteria is difficult to determine from the structure of the organisms because the organisms are so simple. Other methods, such as analyzing cell proteins, DNA sequencing, and RNA sequencing, have been the best method of studying the evolution of bacteria.

The **RNA** in the **ribosomes** has changed more slowly than the genome has. Because of the slow evolution of the rRNA, it is useful for comparing the ancestry of different strains of bacteria. The **rRNA** makes up about 0.3–0.4% of the code of the genome.

The evolutionary changes in the rRNA have occurred in different segments at different times. The rRNA has been referred to as a molecular clock because of the independent rates of change in the sequences.

The segments that have evolved more slowly are used for determining the more distant relations between strains. The segments that have changed more recently will be the same only for closely related strains of organisms.

The rRNA has three **subunits**: 23s, 16s, and 5s. These are the sedimentation rates of the rRNA. The 23s chains are 3300 nucleotides long, the 16s are 1650 nucleotides long, and the 5s have 120 nucleotides in their chains.

The first sequencing work was done with the 5s rRNA. Complete sequencing of the 5s rRNA was used to construct the phylogenetic trees and organize the bacteria into major divisions.

As the analytic equipment became more powerful, the 16s rRNA was completely sequenced. The information from these studies allowed the classifications to be refined. Eventually, sequencing of 23s rRNA may yield more information for classification.

Other work that has been done with rRNA has been to hybridize with single-stranded DNA to determine the extent of the pairing between rRNA from one strain with DNA from another strain of organisms.

Topic Test 3: Use of rRNA Genes in Phylogenic Systems

True/False

1. 5s rRNA has 3300 nucleotides.

2. 16s rRNA has 1650 nucleotides.

3. 23s rRNA has 120 nucleotides.

Multiple Choice

4. rRNAs code makes up what percentage of the total genome?
   a. 5%
   b. 0.5%
   c. 30%
   d. 0.3–0.4%

5. Ribosomal RNA has how many subunits?
   a. 1
   b. 2
   c. 3
   d. 4

6. What is the designation of the rRNA subunit based on?
   a. mol%GC
   b. Sedimentation rate
   c. Amino acid sequence
   d. Nucleotide sequence

Topic Test 3: Answers

1. **False**
2. **True**
3. **False**
4. **d**
5. **c**
6. **b**

TOPIC 4: DOMAINS

KEY POINTS

✓ *What kingdom are bacteria placed in?*

✓ *What are the other groupings for bacteria?*

Bacteria are considered to be in the kingdom **Monera** or **Prokaryota**. This kingdom includes the Eubacteria (true bacteria); Cyanobacteria, formerly known as blue-green algae; and the Archeobacteria.

Bacteria are grouped into four **divisions** based on the presence or absence of the cell wall and its composition. These divisions are the initial grouping from the ***Bergey's Manual of Determinative Bacteriology***, currently in the 9th edition. Bergey's Manual is considered the standard for bacterial taxonomy. It is named for the first chairman of the editorial board, David H. Bergey. The divisions per Bergey's Manual are division I, Gracilicutes; division II, Firmicutes; division III, Tenericutes; and division IV, Mendosicutes.

Organisms in other kingdoms, such as plants and animals, would be subdivided into classes and orders. Bacteria are not because taxonomists cannot agree on how to do this.

Bacteria are grouped into 33 subdivisions called "sections" by the Bergey's Manual, as follows:

- Section 1: Spirochetes
- Section 2: Aerobic/Microaerophilic, Motile, Helical/Vibrioid, Gram-negative Bacteria
- Section 3: Nonmotile (or rarely motile) Gram-negative Curved Bacteria
- Section 4: Gram-negative Aerobic Rods and Cocci
- Section 5: Facultatively Anaerobic Gram-negative Rods
- Section 6: Anaerobic Gram-negative Straight, Curved, and Helical Rods
- Section 7: Dissimilatory Sulfate or Sulfur-reducing Bacteria
- Section 8: Anaerobic Gram-negative Cocci
- Section 9: Rickettsias and Chlamydias
- Section 10: Mycoplasmas
- Section 11: Endosymbionts
- Section 12: Gram-positive Cocci
- Section 13: Endspore-forming Gram-positive Rods and Cocci
- Section 14: Regular Nonsporing Gram-positive Rods
- Section 15: Irregular Nonsporing Gram-positive Rods
- Section 16: Mycobacteria
- Section 17: Nocardioforms
- Section 18: Anoxygenic Phototrophic Bacteria
- Section 19: Oxygenic Photosynthetic Bacteria
- Section 20: Aerobic Chemolithotrophic Bacteria
- Section 21: Budding and Appendaged Bacteria
- Section 22: Sheathed Bacteria
- Section 23: Nonphotosynthetic, Nonfruiting Gliding Bacteria
- Section 24: Fruiting Gliding Bacteria
- Section 25: Archeobacteria
- Section 26: Nocardioform Actinomycetes
- Section 27: Actinomycetes with Multilocular Sporangia
- Section 28: Actinoplanetes
- Section 29: Streptomycetes and Related Organisms
- Section 30: Maduromycetes
- Section 31: Thermomonospora and Related Genera
- Section 32: Thermoactinomycetes
- Section 33: Other Genera

The organisms in this section do not fit into any of the other sections. The genera include *Glycomyces*, *Kibdelosporangium*, *Kitasatospria*, *Saccharothris*, and *Pasteuria*.

Topic Test 4: Domains

True/False

1. Bacteria are in the kingdom Bacteriota.

2. *Cyanobacteria* were formerly known as blue-green algae.

3. The organisms in Section 12 are gram-positive cocci.

Multiple Choice

4. Section 13 of Bergey's Manual covers
   a. endspore-forming gram-positive rods and cocci.
   b. sheathed bacteria.
   c. actinomycetes.

5. Section 6 of Bergey's Manual covers
   a. Acheobacteria.
   b. fruiting and gliding bacteria.
   c. anaerobic gram-negative straight, curved, and helical rods.

Short Answer

6. Why is *Bergey's Manual of Determinative Bacteriology* named for Dr. David Bergey?

Topic Test 4: Answers

1. **False**

2. **True**

3. **True**

4. **a**

5. **c**

6. Dr. David Bergey was the first chairman of the editorial board of the manual.

APPLICATION

There are a variety of methods for ensuring sterility of medical equipment and pharmaceutical products. One of the methods calls for taking samples of the product after sterilization and culturing the samples. This procedure focuses on the cultural characteristics of the organisms. Samples are cultured in trypticase soy broth aerobically and in thioglycolate broth to determine if there are anaerobic or microaerophilic bacteria present in the product after sterilization.

DEMONSTRATION PROBLEM

Fifty people contract food poisoning. Fifteen people live in town A and the other 35 live in town B, 100 miles away. Laboratory testing on specimens from the people in the two towns shows *Salmonella enteritidis*. Serotyping of the *S. enteritidis* from specimens in the two towns shows different serologic types.

Solution

The food poisoning probably was not from the same source.

Chapter Test

True/False

1. Most of the classification of bacteria is based on the phenotype.

2. Phenetic classification is based on the phenotype/genotype.

3. One of the nine characteristics for categorizing an organism is where it was first found.

4. Carbohydrate metabolism is a biochemical characteristic for grouping organisms.

5. One genomic characteristic is the percentage of guanine and cytosine in the total DNA of the organism.

6. If two organisms have the same mol%GC, they are definitely related.

7. tRNA is used for determining the genus and species.

8. There are 33 sections/subdivisions in the 9th edition of *Bergey's Manual of Determinative Bacteriology*.

Multiple Choice

9. Binomial nomenclature uses the names of the
   a. genus.
   b. phyllum.
   c. family.
   d. specific epithet.
   e. a and d

10. Binomial means
    a. genus.
    b. two words.
    c. two kingdoms.
    d. two names.

11. Ribosomal RNA subunits in bacteria are
    a. 80s.
    b. 23s.
    c. 16s.

d. 5s.
e. b, c, d

12. A dendrogram is also called a
    a. dichotomous key.
    b. identification chart.
    c. matching coefficient.

13. Characteristics used for creating dendrograms are
    a. indole production.
    b. hydrogen sulfide production.
    c. phenylalanine deaminase.
    d. nitrate reduction.
    e. All of the above

14. The divisions of bacteria based on their cell wall composition in the 9th edition *Bergey's Manual of Determinative Bacteriology* are
    a. gracilicutes.
    b. firmicutes.
    c. tenericutes.
    d. mendosicutes.
    e. All of the above

15. A dendrogram is a
    a. type of matching coefficient.
    b. tree structure plot of the similarities among strains of organisms tested.
    c. matrix.

Short Answer

16. Why aren't bacteria divided into classes and orders as plants and animals are?

17. What classification work was done from complete sequencing of the 5s rRNA?

18. What classification work was done from complete sequencing of 16s rRNA?

Chapter Test Answers

1. **True**
2. **True**
3. **False**
4. **True**
5. **True**
6. **False**
7. **False**
8. **True**
9. **e**

10. **d**

11. **e**

12. **a**

13. **e**

14. **e**

15. **b**

16. Taxonomists cannot agree on how to divide bacteria into classes and orders.

17. Complete sequencing of the 5s rRNA was used to construct the phylogenetic trees and organize the bacteria into major divisions.

18. The information from these studies allowed the classifications to be refined.

Check Your Performance

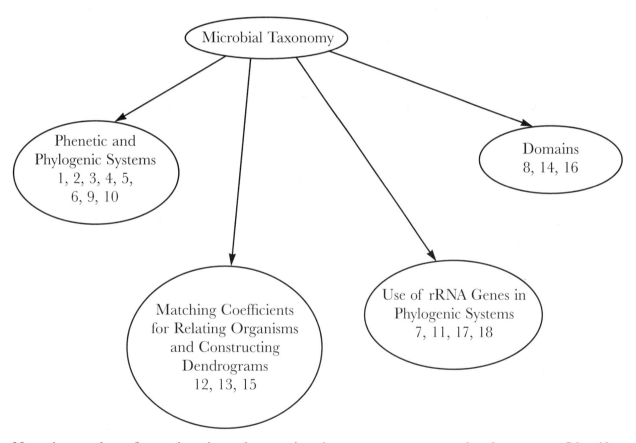

Note the number of questions in each grouping that you got wrong on the chapter test. Identify areas where you need further review and go back to relevant parts of this chapter.

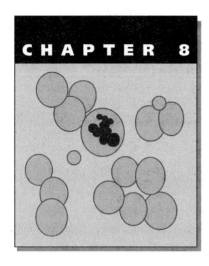

CHAPTER 8

Control of Microorganisms

There is a variety of ways to control the growth of microorganisms. Chemicals, such as isopropyl alcohol, kill the organisms on skin and surfaces. Other chemicals can be taken orally or by injection that will kill or inhibit microorganisms. Also, physical agents, such as UV and ionizing radiation, kill microorganisms.

ESSENTIAL BACKGROUND

- **General knowledge of organic chemistry or biochemistry**
- **General knowledge of microbial cell structures**

TOPIC 1: INHIBITORS OR KILLERS

KEY POINTS

✓ *What is selective toxicity?*

✓ *How does inhibition differ from killing?*

✓ *What are some of the mechanisms of inhibition and killing of microorganisms?*

Antimicrobial agents must be selectively toxic. **Selective toxicity** means that the antimicrobial agent is more toxic to the microorganism than to the host. The ideal antimicrobial agent has *no* effect on the host, whereas it kills or inhibits the infecting organisms.

Some antimicrobial agents come close to the ideal, because they act on structures that exist only in microorganisms, such as murein cell walls and 70s ribosomes. Other useful antimicrobial agents can be toxic to the host. Chloramphenicol is an example because it causes aplastic anemia in a few people. Many other antimicrobial agents fall somewhere between the ideal and the type used as a last resort because they can be toxic to the host. Agents that always kill the host may be useful for sterilizing equipment.

The mode of action of antimicrobial agents can be grouped into one of two categories: killing or *bacteriocidal, viricidal,* or *fungicidal* and inhibition of the pathogen or *bacteriostatic, viristatic,* or *fungistatic.* The "static" and "cidal" compounds are active against specific organisms or types of organisms, so that organism type is used in the category name, such as fungistatic.

Topic Test 1: Inhibitors or Killers

True/False

1. An antimicrobial agent that kills the host and the infecting microorganism has no use at all.

2. Bacteriostatic agents act by killing the pathogen.

3. Antimicrobial agents may be chemicals.

Multiple Choice

4. Selective toxicity means that
   a. the antimicrobial agents only attack certain microorganisms.
   b. agents that sterilize medical instruments.
   c. the antimicrobial agents are more toxic to the microorganisms than to the hosts.

5. Antimicrobial agents may act on cell structures unique to the microorganisms. Examples of these structures are
   a. 70s ribosomes.
   b. murein cell walls.
   c. lysosomes.
   d. endoplasmic reticulum.
   e. a and b

Short Answer

6. If an antimicrobial agent always kills the host, how might it be used?

Topic Test 1: Answers

1. **False.** An antimicrobial agent that kills the host and the infecting microorganism may be useful as a disinfectant or sterilizing agent.

2. **False.** Bacteriostatic agents act by inhibiting the pathogen.

3. **True**

4. **c**

5. **e**

6. It can be used as a sterilizing agent for medical equipment.

TOPIC 2: KINETICS OF KILLING BY AGENTS

KEY POINTS

✓ *What is an antibiotic?*

✓ *How do antimicrobial agents work?*

✓ *How do broad-spectrum antibiotics differ from narrow-spectrum antibiotics?*

Many microorganisms produce substances that kill other microorganisms. We use many of those substances in the treatment of infection. When those toxic substances are used as antimicrobial agents, they are called **antibiotics**. Antibiotics are produced by fungi and by bacteria. Antimicrobial agents that are not synthesized by microorganisms are generally categorized as **chemotherapeutic agents**.

Antimicrobial agents that kill or inhibit both gram-positive and gram-negative organisms are called **broad-spectrum antimicrobial agents**. Some broad spectrum antimicrobial agents are cephalosporins, chloramphenicol, and tetracyclines.

Other antimicrobial agents kill or inhibit only a few taxonomic groups of microorganisms or only gram-positive or only gram-negative organisms. This type of antimicrobial agent is considered to be **narrow spectrum**. Examples of narrow spectrum antimicrobial agents are penicillin G and erythromycin. **Table 8.1** summarizes the modes of action of antimicrobial agents.

Organisms become resistant to the antimicrobial agents in several ways. First, enzymes are developed that destroy the agents. An example is *Staphylococcus aureus* and other penicillin-resistant organisms have developed an enzyme that breaks the beta-lactam ring of penicillin. The enzyme is called beta-lactamase. Second, the organisms may mutate and develop cell membranes that are no longer permeable to the agents. Third, resistant bacteria have altered binding sites. Fourth, organisms may develop alternate metabolic pathways. Fifth, resistance to analogue drugs by developing altered forms of the enzymes that were previously inactivated by the analogues. Finally, some organisms have developed a process that leads to active export of the agent.

In some cases the genes for resistance are part of the bacterial chromosome. Plasmids code for many types of antimicrobial agent resistance. Resistance is transmitted to other organisms, even organisms of different genera or species by conjugation. In conjugation, the plasmid DNA is transferred to the nonresistant organism. Genetic material containing the genes for resistance

Table 8.1. Modes of Action of Antimicrobial Agents

Mode of Action	Example Agent	Active Against	Spectrum
Disruption of cell wall formation and function	Penicillin	Gram positive	Narrow
	Cephalosporins	Gram positive	Narrow
	Vancomycin	Gram positive	Narrow
Alteration of cell membrane function	Polymyxins	Gram negative	Narrow
	Amphotericin B	Fungi	Narrow
	Nystatin	Fungi	Narrow
	Other polyene antibiotics	Fungi	Narrow
Inhibition of protein synthesis by attaching to subunits of 70s ribosomes	Streptomysin	Gram pos/neg	Broad
	Aminoglycoside antibiotics	Gram pos/neg	Broad
	Erythromycin	Gram positive	Narrow
	Chloramphenicol	Gram pos/neg	Broad
	Tetracyclines	Gram pos/neg	Broad
Inhibition of nucleic acid synthesis	Rifampin	Gram pos/neg	Broad
Analogues of nucleic acids that inhibit nucleic acid synthesis	Azidothymidine (AZT)	Virus	
	Ribavirin	Virus	
	Other nucleic acid analogues	Virus	
Inhibition of enzyme function	Sulfa drugs	Gram pos/neg	Broad
Blocking metabolic functions with heavy metals such as arsenic and antimony	Antiparasitic drugs	Parasites	

may also be transferred to cells by phage. Another way that resistance is transferred from one bacterium to another is via transposons. (See Chapter 11, topic 1, for more information on transposons.)

Common methods for testing the susceptibility of bacteria for certain antimicrobial agents are the minimum inhibitory concentration (MIC) test and the Kirby-Bauer test. The MIC consists of serial dilutions of the antimicrobial agent in agar or tubes that are inoculated with the test organism and incubated. The lowest concentration that prevents growth of the organism is the MIC. The Kirby-Bauer test uses filter paper disks that are saturated with various antimicrobial agents. The disks are placed on an agar plate that was seeded with the test organism. After incubation for 16 to 18 hours, the zones of growth inhibition around the disks are measured and compared with the charts to determine the susceptibility of the test organism for each antimicrobial agent tested. Charts are necessary for interpretation of the Kirby-Bauer test, because the zone of inhibition is dependent on how much the antimicrobial agent can diffuse into the agar.

Topic Test 2: Kinetics of Killing by Agents

True/False

1. Penicillin is a broad-spectrum antibiotic.

Multiple Choice

2. Modes of action of antimicrobial agents are to
   a. inhibit cell wall synthesis.
   b. inhibit protein synthesis.
   c. inhibit cell membrane function.
   d. inhibit nucleic acid synthesis.
   e. All of the above

3. Antibiotics are synthesized by some species of
   a. fungi.
   b. bacteria.
   c. viruses.
   d. parasites.
   e. a and b

4. Pathogens become resistant to antimicrobial agents in one or more of the following ways:
   a. Produce enzymes that destroy antimicrobial agents
   b. Change membrane permeability to prevent uptake of antimicrobial agents
   c. Alter metabolic pathways
   d. Alter binding sites of cell structures
   e. All of the above

Short Answer

5. What is the difference between a narrow-spectrum and a broad-spectrum antibiotic?

Topic Test 2: Answers

1. **False.** Penicillin is a narrow-spectrum antibiotic.

2. **e**

3. **e**

4. **e**

5. A broad-spectrum antibiotic is one that is active against both gram-positive and gram-negative organisms and organisms of many different genera.

TOPIC 3: PHYSICAL AGENTS

KEY POINTS

✓ *What is the main use of physical antimicrobial agents?*

✓ *Which agents are widely used?*

There are several **physical antimicrobial agents** in common use. Because the physical agents would harm a human or animal host, they are used for sterilizing medical instruments, pharmaceuticals, and the air in operating rooms. Those agents include the following:

1. Heat: boiling, tyndallization, pasteurization, autoclaving (steam), and dry heat. Heat denatures or oxidizes proteins and nucleic acids in the cells. Note that boiling does *not* kill spores.

2. Ionizing radiation: from ultraviolet light, cobalt-60, and cesium-137 damages DNA and creates superoxides in the cells.

3. Filtration physically removes organisms from liquids and air.

4. pH extremes denature cell proteins. Organic acids used to slow food spoilage.

5. Osmotic pressure uses sugar and salt to dehydrate organisms by decreasing osmotic pressure. Used in food preservation.

Topic Test 3: Physical Agents

True/False

1. The physical antimicrobial agent that commonly uses sugar to kill the cells acts by decreasing the osmotic pressure in the environment.

2. Boiling kills vegetative cells and spores.

Multiple Choice

3. The sterilization method that removes microorganisms rather than killing them is
   a. filtration.
   b. radiation.
   c. boiling.

4. The sterilization method(s) which kill(s) primarily by DNA damage is
   a. dry heat.
   b. boiling.
   c. UV irradiation.
   d. filtration.

5. The three forms of heat sterilization are achieved by
   a. ionizing radiation.
   b. boiling.
   c. dry heat.
   d. saturated steam.
   e. b, c, and d

Short Answer

6. How does osmotic pressure act as a method of sterilization?

Topic Test 3: Answers

1. **True**

2. **False.** Boiling kills only vegetative cells.

3. **a**

4. **c**

5. **e**

6. Osmotic pressure dehydrates the organisms.

TOPIC 4: CHEMICAL AGENTS

KEY POINTS

✓ *How are chemical agents used?*

✓ *Why are chemical agents rarely used internally?*

Chemical agents can be used as **topical antiseptics** for reducing the bacterial count in water, or for **chemical sterilization** of items that cannot withstand heat or that are made of materials that will be altered by ionizing radiation. In general, chemical agents are too toxic to the host cells to be used internally.

Antiseptics are used on skin to reduce the microbial count. **Disinfectants** are used on surfaces for reduction of the numbers of microbes. **Sterilizing agents** kill all the microorganisms on the surfaces of the items sterilized. **Protein denaturation** is a term for damaging the tertiary structure or folding of protein chains. Cooked egg white is a common example of a denatured protein. Information on chemical antiseptics, disinfectants, and sterilizing agents is summarized in **Table 8.2**.

Table 8.2. Categories and Uses of Chemical Antiseptics, Disinfectants, and Sterilizing Agents

CATEGORY OF ACTION	CHEMICALS	USED FOR
Protein denaturation	Phenols	Skin and surface disinfection
	Alcohols	Skin and surface disinfection
Alkylating agents	Formalin	Chemical sterilization
	Glutaraldehyde	Chemical sterilization
	Ethylene oxide	Chemical sterilization
	Propylene oxide	Chemical sterilization
Disruption of disulfide bonds in proteins	Heavy metals	Topical application as skin antiseptics
	Mercury	
	Silver	
	As silver nitrate	Used as eye drops to prevent neonatal eye infection by *N. gonorrhea*
		Used on dressings to prevent burn infections
	Other heavy metals	Skin antiseptics and surface disinfectants
Oxidizing agents	Ozone	Water treatment
	Hydrogen peroxide	Skin antiseptic
	Chlorine/hypochlorite	Water treatment
	Other halogens	Water treatment and skin antiseptics
Disruption of cell membranes	Detergents: Example quaternary ammonium compounds	Surface disinfection
	Surfactants	Surface disinfection
Miscellaneous chemicals used in food preservation	Organic acids	Food preservation
	Nitrites	Food preservation
	Sulfites, sulfur dioxide	Food preservation
	Salt (NaCl)	Food preservation
	Oils from herbs	Food preservation and dental antiseptic

Topic Test 4: Chemical Agents

True/False

1. Heavy metals are used topically.

2. Detergents disrupt the cell walls.

Multiple Choice

3. Ethylene oxide and formalin are examples of
   a. oxidizing agents.
   b. halogens.
   c. heavy metals.
   d. alkylating agents.

Short Answer

4. What is protein denaturation?

Topic Test 4: Answers

1. **True**

2. **False.** Detergents disrupt the cell membranes.

3. **d**

4. Protein denaturation is the disruption of the tertiary structure of proteins—the folding of the proteins into functional form.

APPLICATION

Ethylene oxide (ETO) sterilization is used by manufacturers and hospitals for sterilizing many medical products, specifically those made from plastics, such as syringes, intravenous tubing, dialysis tubing, medicine cups, and anything else that would be damaged by heat or radiation. Other more exotic users of ETO sterilization are museums, which sterilize pelts and animal skins that will be displayed, and NASA, for sterilizing components that will be sent into space, so we do not introduce our bacteria with the equipment we send into space.

Sterilization of medical products is a statistical process. The sterilization parameters needed for a certain product are determined by sterilizing test loads of the product with an indicator organism. The indicator organism for ETO sterilization is the spore of *Bacillus subtilis* var. *niger*. Known quantities of spores are applied to samples of the product or spore strips—filter paper strips with known concentrations of spores are used.

The statistical interpretation is that if the sterilizer cycle will kill 1000 spores, there is a 1 in 1000 chance of a nonsterile item. A cycle that kills 1,000,000 spores has only a 1 in 1 million chance that a nonsterile item will come out of the sterilization process.

Steps in an industrial ETO sterilization process.

1. Preconditioning. At least 24 hours of preconditioning before sterilization is done by putting the product for sterilization into a room that has 90% or greater relative humidity and a temperature of approximately 85°F. This prehumidification allows the ETO gas to penetrate packaging and boxes and ultimately the medical devices more efficiently.

2. Ambient air evacuation. When the product is put into the sterilizer, the first step is for a vacuum to be pulled on the chamber. This removes the ambient air, so the ETO gas will not be diluted when it is pumped into the sterilizer. How much vacuum depends on what is being sterilized. Some items cannot withstand an intense vacuum. (See note at end on ETO gas.)

3. Humidification. The next step in the sterilizer cycle is to pump saturated steam into the chamber. This is a low-pressure process. The pressure in the chamber frequently does not go higher than ambient air pressure—in other words, the steam cycle relieves a part of the vacuum. The product is allowed to sit in the steam environment for an hour or more, depending on the defined parameters for the type of product being sterilized.

4. ETO gas injection. ETO gas is pumped into the sterilizer chamber. How high the gas pressure goes depends on the product inside, again. Some products can withstand up to 15 psi pressure and others cannot; 10–15 psi is a fairly typical pressure range for ETO

sterilization. The operating temperature of an ETO sterilizer is usually about 60°C, but this too will depend on the product being sterilized. (See note at end on ETO gas.)

5. Gas exposure. The ETO cycle will last for several hours. As gas is absorbed by the product, more will be injected to maintain the pressure. The length of the ETO exposure cycle depends on how long it takes to kill the specified number of *B. subtilis* var. *niger* spores in a sterilizer load. The more dense the pallet loads of product, the longer the ETO exposure cycle needed to sterilize the product. Four to 12 hours of gas exposure is common for large loads of product in industrial sterilizers.

6. Air wash. The final step in the sterilization process is the "air wash" cycle. The air wash consists of pulling a vacuum on the sterilizer chamber and then allowing filtered air to enter the chamber. This vacuum, air-in process may be repeated several times to remove as much residual ETO from the product as possible.

It may be necessary to allow a few days for the residual ethylene oxide, ethylene chlorhydrin, and ethylene glycol in the sterilized product to diffuse out of the product.

ETO gas for industrial use is usually diluted with fluorocarbon gas, such as dichlorodifluoromethane or carbon dioxide to keep it from exploding. If the fluorocarbon gas is used, the ratio of ETO to fluorocarbon is usually 12/88. When carbon dioxide is used, the ratio of ETO to CO_2 is usually 10/90.

Some industrial sterilizers use 100% ETO. These sterilizers must have an absolute vacuum pulled at the beginning of the cycle to prevent explosion that can be caused by ETO gas mixing with air. The gas pressure for a 100% ETO sterilizer never gets as high as ambient. Pure ETO is extremely explosive and flammable. It must be diluted with 22 times its volume of water to control an ETO fire.

For more information on ETO sterilizers and the sterility testing, see http://www.namsa.com/ for information on spore strips and sterility testing by a contract lab and http://www.etcusa.com/eto.htm sterilizer photo

DEMONSTRATION PROBLEM

The MIC test uses serial dilutions of the antibiotic. You have four tubes, containing 0.5 mL of saline and one tube with 1 mL of undiluted antibiotic. Transfer 0.5 mL from each tube to the next in the series. What are the dilutions in your series?

Solution:

#1 = 1:1, #2 = 1:2, #3 = 1:4, #4 = 1:8, and #5 = 1:16.

Chapter Test

True/False

1. If an antimicrobial agent is too toxic to the host, it may be useful for sterilization or disinfection of inanimate objects.

2. Silver nitrate drops are used to prevent blindness in newborns from *Neisseria gonorrhoeae*.

3. Ozone and hypochlorite compounds are examples of oxidizing agents.

4. Narrow-spectrum antibiotics are effective against only gram-negative or gram-positive organisms or only a few genera.

5. Infrared is a form of ionizing radiation used to sterilize.

6. Ionizing radiation is used on skin.

7. pH is used as a microbial growth inhibitor.

8. Ethylene oxide is an alkylating agent.

9. The Kirby-Bauer test is a test of resistance to ionizing radiation.

10. Minimum inhibitory concentration is the test of the minimum amount of organic acids needed to slow food spoilage.

11. Many chemical agents work by denaturing cellular protein.

Multiple Choice

12. Plasmids often store the genes for
    a. bacterial cell wall characteristics.
    b. antibiotic resistance.
    c. ribosomes.

13. Sulfa drugs are an example of
    a. metabolic analogues.
    b. penicillins.
    c. tetracyclines.
    d. antiviral agents.

14. What an antimicrobial agent does to the microbial cell is called
    a. mode of action.
    b. cellular disruption.
    c. selective toxicity.

Short Answer

15. What is selective toxicity?

16. Which test of sensitivity to antimicrobial agents uses disks saturated with the agent?

Chapter Test Answers

1. **True**

2. **True**

3. **True**

4. **True**

5. **False**

6. **False**

7. **True**

8. **True**

9. **False**

10. **False**

11. **True**

12. **b**

13. **a**

14. **a**

15. Selective toxicity is when an antimicrobial agent is more toxic to the pathogen than to the host.

16. The Kirby-Bauer test uses disks saturated with the antimicrobial agent.

Check Your Performance

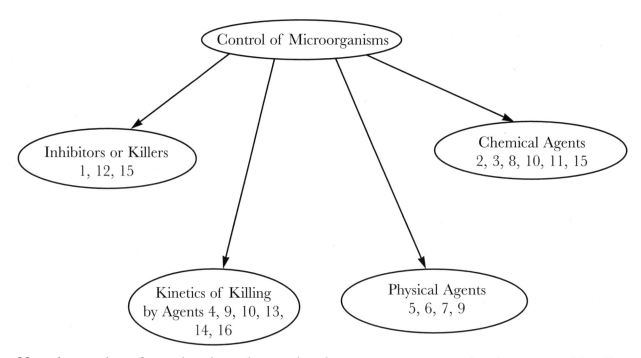

Note the number of questions in each grouping that you got wrong on the chapter test. Identify areas where you need further review and go back to relevant parts of this chapter.

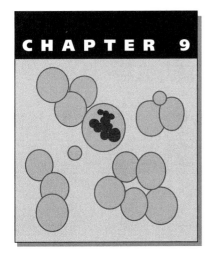

CHAPTER 9

Viruses

Viruses and virus-related particles are the smallest of the known infective agents. They lack the characteristics of cells and because of that, viruses are considered to be *nonliving* entities. Viruses have DNA or RNA but *never* both. The smallest virus is approximately the same size as a 70s bacterial ribosome. Viruses range in size from about 30 nanometers to approximately 300 nanometers. They can be seen only with an electron microscope.

ESSENTIAL BACKGROUND

- **Structure of proteins and nucleic acids**
- **General microbiology**

TOPIC 1: STRUCTURE, SHAPE, AND SPECIFICITY OF VIRUSES

KEY POINTS

✓ *What distinguishes viruses from other microorganisms?*

✓ *What structures make up virus particles?*

✓ *What is specificity?*

The most important characteristic that differentiates viruses from other microorganisms is that viruses have only one type of nucleic acid. A virus has either DNA or RNA but never both.

The DNA or RNA found in viruses may be **single stranded** or **double stranded**. In a few types of viruses the RNA strand is the negative strand and is used as a template for making the positive strand that is needed for virus replication. The nucleic acid in viruses may be linear or circular. It may also be segmented (**Table 9.1**). Viruses do not have independent metabolism. They are obligate intracellular parasites. In the laboratory they are generally grown in embryonating eggs or cell culture, though poliovirus has been grown in ground nonliving human cells.

Because of the differences between viruses and other microorganisms, virus units are called **virus particles** or **virions**, not cells. The differences between viruses and other microorganisms puts them "on the border" between living entities and nonliving entities.

The virus **nucleic acid** or **genome** is surrounded by a capsid, which is a shell composed of protein subunits called capsomeres. The proteins in the capsomeres are determined by the genome. The capsids are specific to the virus. Because the capsids are specific, they are used for

Table 9.1. Viruses vs. Other Microorganisms

	VIRUSES	OTHER MICROORGANISMS
Nucleic acid	DNA or RNA†	DNA and RNA
Have cytoplasm	No	Yes
Have subcellular organelles	No	Yes
Have own enzymes	No*	Yes
Cells divide, grow, metabolize	No	Yes
Have independent metabolism	No	Yes

* HIV does have one enzyme that is inactive until the virus infects the host cell.
† The nucleic acid in viruses may be linear or circular. It may also be segmented.

identification of the virus. The capsids are also generally good antigens because they are proteins and unique to the specific virus.

The capsids may have a variety of shapes, depending on the type of virus:

1. Helical or spiral. The capsid of the Tobacco Mosaic virus is this shape.

2. Polyhedrons, such as icosahedrons, which are composed of 20 triangles. Human adenovirus is one of the polyhedrons.

3. Complex capsid. This type of capsid may be a combination of a polyhedron and a helix. Bacteriophage viruses that infect bacteria have complex capsids.

4. Bullet shaped. Rhabdovirus, the virus that causes rabies is bullet shaped.

The shape of the virus may be spherical if the virus has an **envelope**. The envelope is composed of lipids, proteins, and carbohydrates that have been formed from the membranes of the host cell. The genome codes for the specific chemical structure of the envelopes. The viruses use the membrane bound organelles, such as the Golgi apparatus and endoplasmic reticulum and the nuclear and cytoplasmic membranes for their replication.

Not all viruses have envelopes. Those that do have the envelopes are less susceptible to the host's immune system, because the envelope is similar to host cell membranes and is not recognized as foreign. The envelopes also help the virus attach to the host cell membrane and fuse with the host cell.

The viruses with envelopes are more "fragile" than viruses without envelopes. The viruses with envelopes are susceptible to many antimicrobial agents, such as solvents that dissolve lipids and chemicals such as hydrogen peroxide, chlorine, and phenol. Viruses with envelopes are also more susceptible to extremes of temperature and pH.

Some viruses have **protein spikes** that attach to host cell receptors. The spikes are called **peplomers**. In some cases, the spikes cause clumping of red blood cells—hemagglutination.

Bacteriophage have tail and tail fibers for attachment to host bacterial cells. These structures cause the capsid to be categorized as complex.

Most viruses do not have any enzymes. HIV is one virus that contains an inactive enzyme that becomes active in the host cell for replication of the virus particles. When the virus particles are "packaged," an inactive form of the enzyme is packaged along with the genome.

Virus particles that do not have an envelope are composed of only the capsid and nucleic acid. They are called **nucleocapsids** or naked viruses.

Many viruses are extremely host specific. They may infect only one species or they may only infect certain cells within that species. A few viruses have a wider range of acceptable hosts. Rabies, which can infect warm-blooded animals, is an example of a virus with broad specificity.

Topic Test 1: Structure, Shape, and Specificity of Viruses

True/False

1. Viruses have both DNA and RNA.

2. Virus units are called virus particles, not cells.

3. Viruses with envelopes are more fragile than naked viruses.

4. Viruses have a protein coat called a capsomere that is composed of smaller protein units called capsids.

5. Rabies is an example of a virus that is extremely host specific.

Multiple Choice

6. Virus particles may have the following structures:
   a. Capsid
   b. Genome
   c. Envelope
   d. Cell wall
   e. a, b, and c

7. Viruses that have envelopes are more susceptible than viruses without envelopes to
   a. solvents that dissolve carbohydrates.
   b. antibiotics.
   c. solvents that dissolve lipids.

Short Answer

8. What is a nucleocapsid?

Topic Test 1: Answers

True/False:

1. **False.** Viruses can have DNA *or* RNA, but never both.

2. **True**

3. **True.** They are susceptible to solvents that dissolve lipids.

4. **False.** The smaller units are the capsomeres and the assembled protein coat is the capsid.

5. **False.** It has broad host specificity, because it can infect warm-blooded animals in general.

6. **e**

7. **c**

8. A nucleocapsid is another name for a virus that is composed only of the nucleic acid and protein coat. It may also be called a naked virus.

TOPIC 2: VIRULENT AND TEMPERATE BACTERIOPHAGE: LIFE CYCLES

KEY POINTS

✓ *What are lytic phage?*

✓ *What are lysogenic phage?*

✓ *What is lysogenic conversion?*

Phage or **bacteriophage** are viruses that infect bacteria. Phage are often used for the study of viruses because their hosts are so easy to grow in the lab.

Phage **genomes** are like the genomes of other viruses. Their nucleic acid may be single- or double-stranded RNA or DNA.

Many of the commonly studied phage are named T plus a number, such as T4. The "T" stands for "type." The most frequently studied T phage is T4 coliphage, which infects *Escherichia coli*. T4 phages have a complex capsid that has a polyhedron head and a helical tail with tail fibers.

T4 phage is an example of a lytic or virulent phage. Lytic phage replicates in a five-step cycle that ends with the lysis of the infected cell:

1. Attachment or adsorption;

2. Penetration of the host cell;

3. Biosynthesis;

4. Maturation or assembly of the phage;

5. Release of the virus.

Lysogenic or **temperate phage** may not lyse the host cell. Most bacteriophage are lysogenic. Lambda phage is an example of a temperate coliphage, a phage that infects *E. coli*.

The steps in the replication of a temperate or lysogenic phage follow:

1. Adsorption and penetration of the host cell. The DNA is in a linear form at this stage.

2. Integration of the phage genome with the host chromosome. The phage genome is inserted into the host chromosome at a specific site. The integrated phage genome is called **prophage**. Bacterial DNA with the integrated prophage is called a **lysogen**.

3. Replication of the bacterial DNA with the integrated prophage in preparation for bacterial cell division.

4. Host cell division yields two bacteria and two phage integrated onto the host cell genome. Lysogenic prophage integrated on the host cell chromosome changes the genetic characteristic of the host cells. Prophage genes integrated with the host genome cause the host cells to produce proteins that inhibit virus replication and prevent other phage with the

same nucleic acid from infecting the host cell. The prophage does not prevent other types of lysogenic or lytic phage from infecting the host.

5. The genetic alteration of the host by the prophage is called **lysogenic conversion**. Toxin production by *Corynebacterium diphtheria* and *Clostridium botulinum* are the result of lysogenic conversion. If the bacteria are not infected by lysogenic phage, they do not produce their toxins.

6. On occasion, lysogenic phage become lytic. The process is called **induction**. During induction, the phage genome separates from the host genome and begins coding for phage proteins. Once the lytic cycle is induced, the phage replication is the same as lytic phage and ends with lysis of the host cell.

Induction may occur spontaneously. Induction may also be the result of an environmental trigger, such as the lack of nutrients.

Topic Test 2: Virulent and Temperate Bacteriophage: Life Cycles

True/False

1. Lysogenic conversion is responsible for the toxin production by *Clostridium botulinum*.

2. A lysogen is the host genome with the viral genome integrated.

3. T4 is an example of a lytic phage that infects *Escherichia coli*.

Multiple Choice

4. Which of the following steps are part of the lytic cycle:
   a. Integration of the phage genome onto the host chromosome
   b. Penetration of the host cell
   c. Host cell division
   d. Assembly of virus components
   e. b and d

5. The prophage is
   a. a lytic phage before maturation.
   b. a T4 phage.
   c. a phage genome integrated with host DNA.

6. T4 uses _____ to weaken the host cell wall to gain entry into the host:
   a. lysozyme
   b. host cell enzymes
   c. tail of the capsid

Topic Test 2: Answers

1. **True.** The prophage alters the genetic makeup of the host organism to enable it to produce the toxin.

2. **True**

3. **True**

4. **e**

5. **c**

6. **a**

TOPIC 3: ANIMAL VIRUSES

KEY POINTS

✓ *How do animal viruses replicate?*

✓ *How are animal viruses classified?*

Animal viruses go through the same replication steps as lytic phage, but the process that occurs at those steps is different. The steps in the replication of animal viruses are as follows:

1. Adsorption.

2. Penetration is often by engulfment by host.

3. Biosynthesis.
   a. Replication of viral DNA is often done in the host cell nucleus by viral enzymes.
   b. Virus RNA synthesis has more variations depending on whether the RNA is double-stranded, single-stranded positive sense, or single-stranded negative sense.

4. Maturation or assembly of the viral components takes place most often in host nucleus, cytoplasm, and at cytoplasmic membrane.

5. Release of the virus may or may not lyse the host cell. Viruses that bud through the cell membrane without lysing cell get envelopes made of cell membrane.

Laboratory culture of animal viruses is done in embryonating eggs or cultures of animal or human tissue cells. Plaque assays for virus counts can be done in cell culture. These plaque assays are similar to the plaque assays done on phage.

There are three types of cell cultures:

1. Primary cell cultures. These are cells directly from the animal. They have not been subcultured.

2. Continuous cell lines. These cells have been subcultured for many generations, such as the HeLa cells, which have been subcultured since 1951.

3. Diploid fibroblasts are cells from fetal tissue. They are used because they are rapidly dividing and growing. They are usually free of contaminating viruses. Diploid fibroblasts produce collagen and other fibrous connective tissue components.

 Classification of animal viruses is based on several criteria. The main criteria are size of the virus, morphology, DNA/RNA, capsid, and envelope presence. **Table 9.2** shown below gives some of the classification characteristics for the animal virus families. See Chapter 17 for information on the animal viruses and the diseases they cause.

Table 9.2. Animal Viruses

Virus Family Name	Nucleic Acid*	Linear/Circular/ Segmented	Ether Sens/ Envelope	Capsid Size Sm/Med/Lg	Capsid Shape
Parvoviridae	SS DNA	Linear	No/no	Small	Icosahedron
Papovaviridae	DS DNA	Circular	No/no	Small	Icosahedron
Adenoviridae	DS DNA	Linear	No/no	Med	Icosahedron
Herpesviridae	DS DNA	Linear	Yes/yes	Small	Icosahedron
Hepadnaviridae	DS DNA	Circular	Yes/yes	Small	Icosahedron
Poxviridae	DS DNA	Linear	No/yes	Large	Complex
Picornaviridae	SS+RNA	Linear	No/no	Small	Icosahedron
Astroviridae	SS+RNA	Linear	No/no	Small	Icosahedron
Calciviridae	SS+RNA	Linear	No/no	Small	Icosahedron
Togaviridae	SS+RNA	Linear	Yes/yes	Small-med	Icosahedron
Flaviridae	SS+RNA	Linear	Yes/yes	Small	Unknown
Arenaviridae	SS+RNA	Segmented	Yes/yes	Med-large	Complex
Coronaviridae	SS+RNA	Linear	Yes/yes	Med-large	Helical
Retroviridae	SS+RNA	Diploid[†]	Yes/yes	Med-large	Helical
Bunyaviridae	SS−RNA	Circular	Yes/yes	Med	Helical
Orthomyxoviridae	SS−RNA	Linear Segmented	Yes/yes	Med	Helical
Paramyxoviridae	SS−RNA	Linear	Yes/yes	Large	Helical
Rhabdoviridae	SS−RNA	Linear	Yes/yes	Med	Complex-helical
Filoviridae	SS−RNA	Linear	Yes/yes	Med-large	Pleomorphic
Reoviridae	DS RNA	Linear Segmented	No/no	Med	Icosahedron

* SS, single stranded; DS, double stranded; +RNA, positive sense; −RNA, negative sense.
† There are 2 copies of the viral RNA in the virus genome.

Topic Test 3: Animal Viruses

True/False

1. *Parvovirus* is the only single-stranded DNA animal virus.

2. *Reovirus* is the only known double-stranded RNA animal virus.

3. *Picornavirus* gets its name from pico meaning small and RNA.

Multiple Choice

4. Possible configurations of virus genomes are
   a. double-stranded DNA.
   b. single-stranded DNA.
   c. double-stranded RNA.
   d. single-stranded positive sense DNA.
   e. single-stranded negative sense RNA.
   f. All except d

5. Which are DNA viruses?
   a. *Parvoviridae*
   b. *Flaviviridae*
   c. *Hepadnaviridae*
   d. *Astroviridae*
   e. a and c

6. Some criteria for animal virus classification are
   a. DNA/RNA.
   b. capsid shape.
   c. morphology.
   d. size.
   e. All of the above

Topic Test 3: Answers

1. **True**
2. **True**
3. **True**
4. **f**
5. **e**
6. **e**

TOPIC 4: VIRINOS, VIROIDS, PRIONS, AND OTHER SUBVIRAL PARTICLES

KEY POINTS

✓ *What are prions and what diseases do they cause?*

✓ *What are the hosts of viroids?*

Viroids are small circular pieces of single-stranded RNA. Viroids replicate independently of other viruses. They do not have protein coats or code for proteins. They use host enzymes for nucleic acid synthesis. Viroids are the smallest known infective agent. Viroids are believed to have evolved independently from viruses.

Virinos, also called **prions**, are proteins. Prions are highly infective. Prions have no detectable nucleic acid. They are sensitive to protease enzymes but not to nuclease enzymes. Prions are resistant to many common antimicrobial agents, such as formalin, ethanol, deoxycholate, ionizing radiation, proteases, and *beta*-propionolactone.

Prions can be inactivated by 90% phenol, household bleach, ether, acetone, urea, strong detergents, iodine disinfectants, and autoclaving. For skin and instrument disinfection, guanidine thiocyanate is a good choice.

Table 9.3 shows the subviral particles and their characteristics. **Table 9.4** shows the characteristics of prion diseases. **Table 9.5** lists prion diseases and their hosts.

Table 9.3. Subviral Particles				
SUBVIRAL PARTICLE	NUCLEIC ACID	REQUIRES HELPER FOR REPLICATION	PROTEIN COAT Y/N	HOST
Satellite virus	SS RNA	Yes	Yes	Plants
Satellite RNA	SS RNA	Yes	Yes	Plants
Viroids	SS RNA	No	No	Plants
Prions or virinos	None detectable	No	Yes	Animals

Table 9.4. Prion Disease Characteristics
✓ Attack central nervous system
✓ Cause cavities in neurons
✓ Cause enlargement of the astroglial cells
✓ Cause spongelike changes in the gray matter
✓ Frequently cause amyloid plaques
✓ Long incubation period (up to decades)
✓ Period of increasing deterioration
✓ Always fatal
✓ Never any remission or recovery from the diseases
✓ Diseases are not affected by immunosuppression
✓ No antibody production
✓ Not affected by the cellular immune components
✓ No known treatment

Table 9.5. Prion Diseases

DISEASE	HOST
Scrapie	Sheep
Transmissible mink encephalopathy	Mink
Bovine spongiform encephalitis	Cattle
Kuru	Humans
Creutzfeldt-Jakob (CJ) disease	Humans
Variants of CJ	
Gerstmann-Stradussler-Scheinker	Humans
Fatal familial insomnia	Humans

Topic Test 4: Virinos, Viroids, Prions, and Other Subviral Particles

True/False

1. Viroids infect animal cells.

2. Viroids require a helper virus to replicate.

3. Virino is another name for prion.

Multiple Choice

4. Prions are
   a. protein.
   b. viruses that need helpers to replicate.
   c. bare nucleic acid.

5. Prion diseases have what characteristics in common?
   a. They affect central nervous system.
   b. They are degenerative.
   c. They have no known treatment.
   d. All of the above

6. Some of the diseases caused by prions are
   a. scrapie.
   b. transmissible mink encephalopathy.
   c. bovine spongiform encephalitis.
   d. Creutzfeldt-Jakob disease.
   e. All of the above

Topic Test 4: Answers

1. **False.** Viroids only infect plants.
2. **False.** Viroids replicate independently.
3. **True**
4. **a**
5. **d**
6. **e**

APPLICATION

Many viruses are specific for a single host species or a narrow range of host species. This fact is being exploited in the use of viruses as pesticides. One of the advantages of using viruses as pesticides include their short "life" span. Viral pesticides are active for a short time, then are destroyed by sunlight instead of leaving residues that may never deteriorate like some chemical pesticides. The viral pesticides are so specific they kill off the target pest without harming the desirable insects, such as honey bees.

DEMONSTRATION PROBLEM

If you want to identify a virus found in a patient's blood, what can be determined from electron microscopy?

Solution

Electron microscopy will show the shape of the virus capsid and the size of the capsid for use in classifying the virus.

Chapter Test

True/False

1. When the presence of phage changes the genetic characteristics of the host bacterium, this is called lysogenic conversion.
2. Mad Cow disease is caused by an arbovirus.

3. Prions cause hemorrhagic fevers.

4. Animal viruses go through the same replication steps as lysogenic phage.

5. Phages infect bacteria.

6. Animal viruses get their envelopes from lysing the host cell.

7. Viroids are agents that only infect plants.

Multiple Choice

8. An important characteristic of all viruses is that
   a. they have DNA *or* RNA, never both.
   b. they only infect plants.
   c. they produce toxins.

9. Lytic phage
   a. replicate with host bacterium.
   b. lyse host cell.
   c. cause host bacteria to produce toxins.

10. Virions are
    a. bare nucleic acid.
    b. assembled virus particles.
    c. protein.

Short Answer

11. Where do animal viruses replicate their nucleic acid?

12. Where do animal viruses assemble the completed virion?

13. What is induction?

14. What characteristic of the nucleic acid composition identifies a virus?

15. Why are phage often studied instead of animal viruses?

Chapter Test Answers

1. **True**

2. **False**

3. **False**

4. **False**

5. **True**

6. **False**

7. **True**

8. **a**

9. **b**

10. **b**

11. Animal viruses generally replicate their nucleic acid in the host cell nucleus.

12. Maturation or assembly of the virion takes place at different locations within the host cell, depending on the type of virus. The common sites of assembly are the nucleus of the host cell, the inner surface of the cytoplasmic membrane, and in the cytoplasm.

13. Induction is the process of lysogenic phage becoming lytic. The phage genome disengages from the host genome and takes over the host metabolism to make phage protein and nucleic acid.

14. Viruses have only DNA or RNA. They never have both.

15. Because phages are bacterial viruses, their hosts are easy.

Check Your Performance

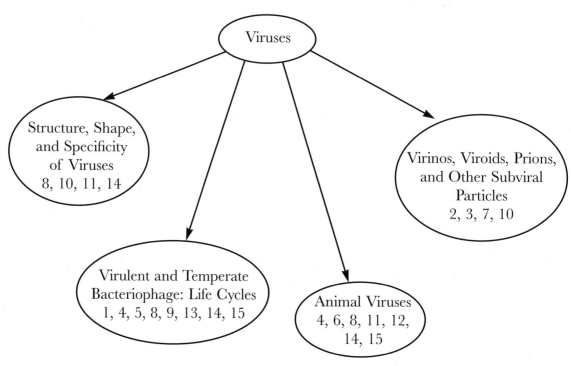

Note the number of questions in each grouping that you got wrong on the chapter test. Identify areas where you need further review and go back to relevant parts of this chapter.

UNIT II
MICROBIAL GENETICS

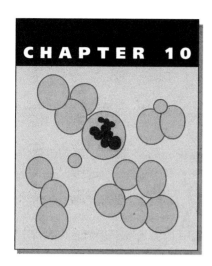

CHAPTER 10

Microbial Genetics: Reproduction and Metabolic Regulation

Just like the cells of eukaryotic organisms, the cells of prokaryotic organisms must store biologic information, replicate that information, and determine which genes in that information should be activated under different environmental conditions. DNA is used by prokaryotes as the informational molecule. Viruses and viroids, whose classification is debatable (Chapter 1, Topic 2), may use single-stranded DNA, single-stranded RNA, or double-stranded RNA to store information in addition to double-stranded DNA.

ESSENTIAL BACKGROUND

- Structure and components of DNA and RNA (Chapter 2, Topic 3)
- Differences between prokaryotic and eukaryotic cells (Chapter 1, Topic 2)

TOPIC 1: REPLICATION, TRANSCRIPTION, AND TRANSLATION IN PROKARYOTES

KEY POINTS

✓ *How are nucleic acids replicated to ensure conservation of their information?*

✓ *How is the information in a gene transcribed into a message?*

✓ *What is produced when an mRNA is translated?*

Both DNA and RNA are replicated using a **semiconservative pattern** and must be double stranded for replication to begin (**Figure 10.1**). Single-stranded DNA and RNA are converted first to a double-stranded replicative form before replication occurs (**Figure 10.2**). Viroids, although single stranded, apparently form intrastrand base pairing, resulting in a double-stranded structure that host cells mistake for double-stranded DNA. Each parent strand of the double-stranded structure serves as a template, guiding the synthesis of a complementary progeny strand. The end product is two nucleic acid molecules containing a conserved parent strand and a new complementary progeny strand, the **semiconservative pattern** (Figure 10.1). Replication of a bacterial chromosome begins at a unique DNA sequence, the **origin**. At this point **DNA polymerase** binds and begins to synthesize new DNA, producing a **leading strand** and a **lagging strand**, in one of two ways: by having the replication point

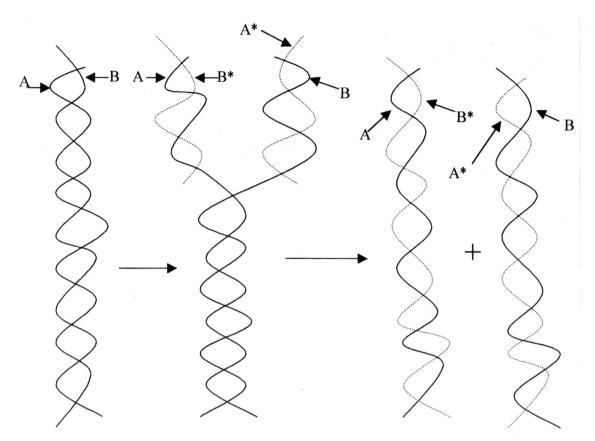

Figure 10.1. Semiconservative replication of DNA.

Figure 10.2. Formation of the replicative form by organisms having single-stranded RNA or DNA.

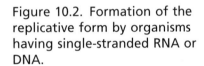

single-stranded
nucleic acid

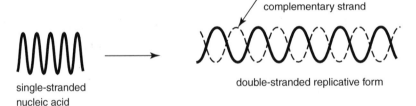

newly synthesized
complementary strand

double-stranded replicative form

move in either direction from the starting point, the **bidirectional model**, or by opening one strand of the circular chromosome and revolving the closed strand, the **rolling circle model**.

Structural genes in DNA contain information for the production of a protein. Some regions of DNA do not code for proteins but serve regulatory functions (see Topic 2).

Transcription begins with the binding of sigma factor of RNA transcriptase (DNA dependent RNA polymerase) to the promoter region of the gene to be transcribed (**Figure 10.3**). RNA transcriptase progresses down the DNA sense strand, and sigma factor drops away. The ribonucleotide sequence of the RNA molecule synthesized is complementary to the DNA sense strand. RNA synthesis continues until an inverted repeat rich in G-C nucleotides is incorporated into the RNA. An intrastrand stem and loop structure forms in the RNA and triggers termination of RNA synthesis. Alternatively, in certain organisms, rho protein binds to the termination

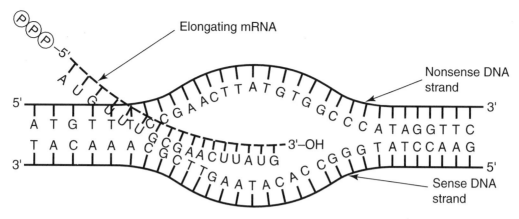

Figure 10.3. Transcription of messenger RNA from a DNA strand.

sequences on the DNA sense strand and blocks further RNA transcriptase activity. Because bacteria have no nuclear membrane, translation of this RNA (**mRNA**) can begin while it still is being synthesized.

Translation involves protein synthesis beginning with binding of mRNA to the small subunit of the ribosome guided by three initiation protein factors and driven by GTP (**Figure 10.4**). While this is occurring, transfer RNAs (**tRNAs**) are activated by the attachment of a particular **amino acid** to the **tRNA** (Figure 10.4a). Each **tRNA** is highly specific for the amino acid that is attached and requires a specific aminoacyl tRNA synthetase to catalyze the attachment. The **codons** (code words) consist of three nucleotides arranged in a continuous sequence of nucleotides on the mRNA. Next, the large ribosomal subunit binds to the complex, providing the P site for the initial aminoacyl-tRNA (Figure 10.4b). The second aminoacyl-tRNA for the developing protein will bind to the A site adjacent to the P site. The amino acid on the tRNA in the P site joins the amino acid on the tRNA in the A site (Figure 10.4b). The **tRNA** in the P site with the help of an elongation factor leaves to be **recharged** with another of its specific amino acid. The ribosome shifts down the **mRNA** one codon aided by two other elongation factors and energy of GTP hydrolysis. The remaining bound **tRNA** is now in the P site, leaving the A site open to receive the aminoacyl-tRNA carrying the third amino acid in the growing **protein** chain (Figure 10.4b). After peptide bond synthesis, the process repeats itself until a stop codon is encountered. Releasing factors bound to the stop codon trigger the ribosomal complex to disintegrate, releasing a completed **protein**. Bacterial mRNAs typically have 20 ribosomes translating simultaneously, forming a polyribosome. Messenger RNAs have about a 5-minute half-life before they are digested by RNases. Having mRNAs with short duration allows cells to change which mRNAs are transcribed in response to changes in the cellular environment.

Topic Test 1: Replication, Transcription, and Translation in Prokaryotes

True/False

1. Semiconservative replication of nucleic acid is used by all nucleic acid containing organisms.

2. Transfer RNA molecules are used only once.

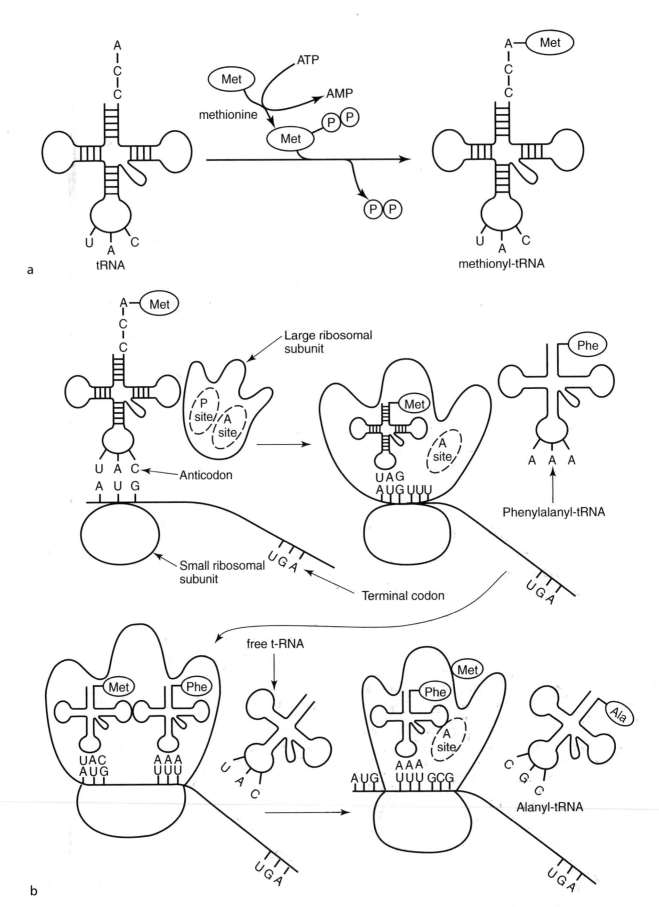

Figure 10.4a. Charging of transfer RNA^{met} with methionine to form methionyl-tRNA^{met}.
b. Translation of messenger RNA by transfer RNAs on the ribosome to synthesize a protein.

Multiple Choice

3. RNA transcriptase finds the initiation site for messenger RNA synthesis by
   a. finding the nucleotide sequence AUG.
   b. attaching to tRNA molecules whose anticodon binds to a promoter site.
   c. attached sigma factor binds to promoter site.
   d. attaching to rho factor.
   e. binding of initiation factors to DNA operator and release of energy in GTP.

4. The energy to institute initiation and elongation is provided by
   a. GTP.
   b. NADH.
   c. ATP.
   d. acetyl-CoA.
   e. pyrophosphate.

Short Answer

5. Why do prokaryotic messages (mRNA) have a half-life of less than 5 minutes?

6. How are the ends of mRNA molecules determined during their synthesis?

Topic Test 1: Answers

1. **True.** The conserved strand serves as a template on which a new complementary strand can be constructed.

2. **False.** Transfer RNA molecules return repeatedly to the cytoplasm to be "recharged" with their specific amino acid.

3. **c.** Although sigma factor leaves the core RNA transcriptase enzyme after mRNA synthesis has begun, the core enzyme cannot bind to DNA without sigma factor being present.

4. **a.** For unknown reasons energy for transformation is provided by the conversion of GTP to GDP and inorganic phosphate. ATP provides energy for "charging" tRNAs with their respective amino acids. The other molecules named do not provide energy to any step in protein synthesis.

5. Short life span of mRNA molecules permits cells to respond quickly to environmental changes that might necessitate different proteins for survival. The cell, thus, conserves resources and energy to produce only the substances needed for survival.

6. There are two known mechanisms. One involves the rho protein that recognizes the terminal sequences of the gene and binds interrupting further progression of RNA transcriptase. The other results from the presence of a palindrome in the growing mRNA strand that forms an intrastrand stem-and-loop configuration in the mRNA that disrupts further elongation of the mRNA.

TOPIC 2: CONTROL OF EXPRESSION AT THE TRANSCRIPTIONAL LEVEL

KEY POINTS

✓ *How are degradative pathways regulated by gene induction?*

✓ *How are biosynthetic pathways regulated by gene repression?*

✓ *What is attenuation of biosynthetic pathways?*

✓ *How does CAP protein serve as an activator of an inducible pathway?*

✓ *Is sigma factor important in gene regulation?*

The enzymes to catalyze certain metabolic reactions, such as those for the degradation of glucose, must be available continuously. Their synthesis is **constitutive**; **no** regulation is used.

Prokaryotes can become more energy efficient by synthesizing only the enzymes for those pathways needed in certain environmental conditions. Degradation of lactose, a sugar found only in mammalian milk, represents one of those conditions. Hence, the enzymes used in lactose degradation should be synthesized only when the prokaryote has this sugar available. Synthesis of these enzymes is blocked most effectively at the transcriptional level, establishing a **negative feedback loop**. Frequently, in prokaryotes the genes coding for enzymes in a biochemical pathway are **clustered** and **coordinately controlled** in a unit known as an **operon**, and the resulting mRNA is **polycistronic** (**Figure 10.5a**). When considering a **degradative** process, **gene induction** is used. A **regulator gene** (I gene), not necessarily found on the prokaryote chromosome near the **operon** regulated, codes for a **regulator protein**. The **regulator protein** binds to the **operator** region immediately preceding the **structural genes** of the **operon** in the absence of the compound to be degraded (Figure 10.5a). RNA transcriptase bound to the **promoter** region immediately upstream from the blocked operator cannot proceed through the **regulator protein** to begin transcribing the **structural genes**. **Regulator proteins** have two sites on them: one for binding to the **operator** region on the DNA and a second **allosteric site** for binding to the **inducer**, usually the compound to be degraded or a structurally very similar derivative. Upon binding the **inducer**, the **regulatory protein** undergoes a change in folding, preventing its binding to the **operator** region and freeing RNA transcriptase to transcribe the **structural genes** (**Figure 10.5b**). As the **inducer**, the compound degraded, is consumed, the **operon** returns to its previous blocked state.

Biosynthesis of a compound requires a somewhat different approach. The levels of the **last** compound in the pathway are significant. In this case when there is an abundance of the end product, the cell needs to **repress** further transcription of structural genes needed for the biosynthetic enzymes. Biosynthesis of the amino acid arginine provides an example of **gene repression**. Structural genes coding for biosynthetic enzymes are usually **clustered** and **coordinately controlled** in an **operon** in prokaryotes (**Figure 10.6a**). The **regulator gene** (I gene) of a **gene repression** system codes for an **aporepressor protein** that cannot bind to the **operator** region on the DNA. When there is surplus end product, arginine in our example, the end product will bind to an **allosteric site** on the **aporepressor**, inducing a conformational change in the **aporepressor**. The resulting complex will bind to the **operator**, blocking transcription of structural genes coding for the biosynthetic enzymes (**Figure 10.6b**).

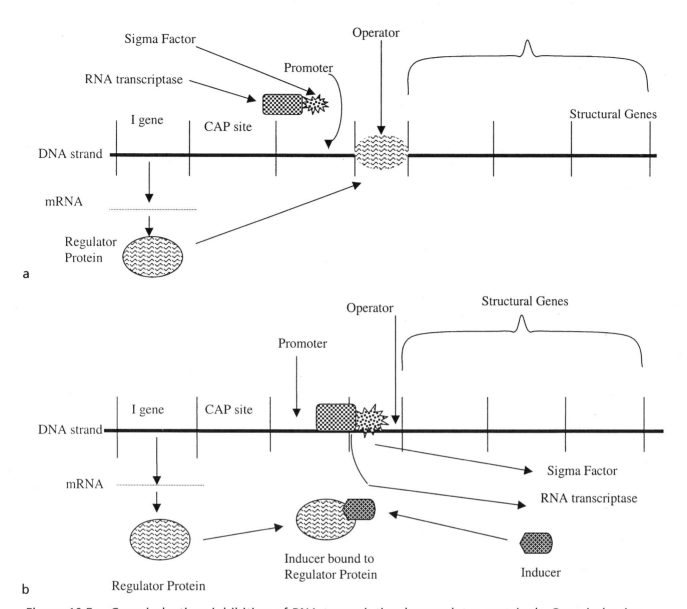

a

b

Figure 10.5a. Gene induction: inhibition of RNA transcription by regulator protein. b. Gene induction.

As a result, the end product of the biosynthetic pathway will be synthesized until its concentration is in surplus to bind the **aporepressor** and shut down further transcription of biosynthetic genes. This represents another form of **negative feedback** loop.

RNA transcription begins with the binding of sigma factor of RNA transcriptase to DNA **promoter** regions. Cells have more than one sigma factor. One alternate sigma factor is associated with heat stress. At the usual culture temperatures cells produce the heat stress sigma factor, but a second protein that has dual functions quickly degrades it. This second protein has **proteolytic** activity on the heat stress sigma factor, but at elevated temperatures is a **chaperonin**, reducing the destruction of heat stress sigma factor. Increased concentrations of heat stress sigma factor results in the increased transcription of heat stress genes, one of which is the second protein. This leads to increased synthesis of the second protein whose **proteolytic** activity decreases heat stress sigma factor and reduces transcription of all genes coding for heat stress proteins again.

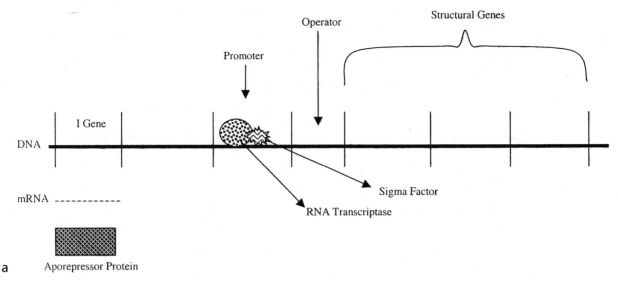

a

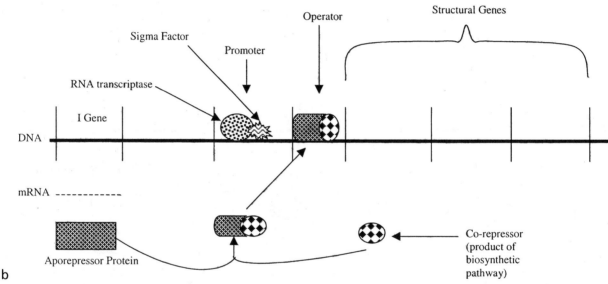

b

Figure 10.6a. Gene repression system of an operon in a biosynthetic pathway, such as amino acid synthesis, in the unrepressed state. b. Gene repression system of an operon in a biosynthetic pathway, such as amino acid synthesis, in the repressed state.

Topic Test 2: Control of Expression at the Transcriptional Level

True/False

1. Synthesis of the enzymes needed for degrading a compound is regulated by the end products of the pathway.

2. The ability of RNA transcriptase to read through an operator can determine the frequency with which the structural genes after that operator are transcribed.

3. Gene repression occurs only in biosynthetic pathways.

Multiple Choice

4. In gene induction what compound initiates the induction process?
   a. mRNA
   b. First compound in the pathway
   c. Regulator protein
   d. Promoter
   e. End product of the pathway

5. The product of the regulator gene in a gene repression system is an aporepressor because
   a. it would be bound continuously to an operator.
   b. it must complex to the end product to bind to an operator.
   c. it must be converted to the active DNA-binding form by proteases.
   d. it is degraded rapidly.
   e. it complexes to the true repressor, and the complex binds to the CAP site, enhancing transcription.

Topic Test 2: Answers

1. **False.** Nothing would happen. Regulation of degradation must be by the first compound in the pathway.

2. **True.** If RNA transcription is blocked, structural genes will not be transcribed.

3. **True.** If gene repression occurred in a degradative pathway, there would be no way of inactivating the pathway. In gene repression, an end product complexing to an aporepressor protein binds to the operator, halting RNA transcription. But in a degradative pathway, end products are present only when substrates are available. Hence, the enzymes of a degradative pathway would be synthesized in the absence of a substrate and reduced when the substrate was present.

4. **b.** The first compound in a degradative pathway complexes with the regulatory protein inhibiting its binding to the operator. RNA transcriptase is free then to progress through the structural genes of the operon.

5. **b.** In a gene repression system if the regulatory protein bound directly to the operator of an operon used in biosynthesis, that operon could never be activated.

TOPIC 3: POSTTRANSCRIPTIONAL CONTROL: CONTROL OF EXPRESSION AT THE TRANSLATIONAL LEVEL

KEY POINTS

✓ *Is regulation of translation as efficient as regulation at the gene level?*

✓ *How is differential binding to start codons achieved?*

Instituting regulation after transcription, but before translation, is not as energy efficient as preventing transcription, that is regulating at the gene level. When posttranscriptional regulation is used, energy is consumed in synthesizing an mRNA molecule but not in translating its message.

Prokaryotic cells have little opportunity to regulate at the posttranscriptional level because prokaryotic messages are being translated before mRNA transcription is complete. Most bacterial mRNAs are polycistronic and are translated sequentially 5′ to 3′ through the cistrons on the mRNA. Between the cistrons on the polycistronic mRNA, there are unread intercistronic sequences that can be up to 40 bases long. The ribosome spans approximately 35 bases on mRNA. The ribosome is large enough to span shorter intercistronic gaps from the termination codon to the subsequent initiation codon. These ribosomes can begin translation of the next cistron without leaving the mRNA. On the other hand, if the intercistronic gap is too large, the ribosome falls off the mRNA at the termination codon and then must bind again to the mRNA at the next initiation codon. Some ribosomes will fail to bind, reducing the number of proteins produced from translation of the following cistron.

RNA bacteriophages have a very precise order in which the cistrons must be read. Secondary folding of the RNA blocks the initiation site of later cistrons. An early cistron whose initiation site is open is translated first. As the first cistron is translated, the ribosomes progressing down the RNA break complementary pair bonding of sequences, binding to the initiation sites of subsequent cistrons. This mechanism ensures an early cistron will be translated before a later one.

Topic Test 3: Posttranscriptional Control: Control of Expression at the Translational Level

True/False

1. Control at the translational level saves as much energy as transcriptional control.

2. Secondary folding of RNA can block the translation of later cistrons.

Multiple Choice

3. The timing of translation of certain cistrons of RNA phages is controlled by
   a. the rate of ribosome attachment to the initial start codon.
   b. the size of ribosome.
   c. ability of ribosome to elongate across an intercistronic gap.
   d. secondary folding of phage RNA so that a complementary sequence blocks a start codon.
   e. None of the above.

Short Answer

4. How can control of the frequency of translation of the later cistrons in polycistronic mRNA be achieved?

Topic Test 3: Answers

1. **False.** Control at the translational level permits energy to be consumed, making mRNAs that are not used.

2. **True.** By blocking start codons of particular cistrons, ribosomes cannot bind to the initiation sites for those cistrons and translation of those cistrons will not occur.

3. **d.** If an initiation site is blocked, translation of the blocked cistron will occur only after ribosomes have moved down the RNA and bound the complementary sequences to the initiation codon.

4. The varying length of intercistronic mRNA sequences affects the ability of a ribosome to remain attached to the mRNA after completing translation of a cistron. If the intercistronic distance is short, the ribosome can span the gap between the terminal codon of one cistron and the initial codon of the following cistron. On the other hand, if the intercistronic distance is long, the ribosome will complete a cistron and then detach from mRNA. Some ribosomes, but not every ribosome, will reattach at the initial codon of the next cistron. Hence, the frequency of translation of that cistron will be less.

TOPIC 4: POSTTRANSLATIONAL CONTROL: CONTROL OF ENZYME ACTIVITY

KEY POINTS

✓ *When is regulation at this level useful?*

✓ *What kind of metabolic pathways would this form of regulation control?*

✓ *How do enzymes with allosteric sites achieve metabolic control?*

Of the three forms of regulation presented in this chapter, **posttranslational control** is the least efficient, saving only the energy expended in the operation of a biochemical pathway. Energy is expended in synthesizing the mRNAs and enzymes needed to perform the reactions in the pathway. **Posttranslational control** affords some degree of regulation over biochemical pathways. These pathways are ones cells cannot afford to inactivate for the approximately 5 minutes required to reactivate gene expression. **Posttranslational control** can be used to provide a second layer of controls over a biosynthetic pathway, such that **posttranslational control** is imposed when amounts of the product of the pathway is somewhat abundant and **transcriptional control** of the genes involved is used when the product is in large excess. The syntheses of aromatic amino acids and vitamins are interrelated and demonstrate both aspects of **posttranslational control** (**Figure 10.7**).

Posttranslational control is known also as **feedback inhibition**. In this method of regulation, the end product of a biochemical pathway binds to the first enzyme in the pathway and interrupts the catalysis of the first reaction. Operation of the pathway halts immediately because no intermediates in the pathway are available. Consumption of the end product releases the initial enzyme to be catalytic again, restarting the pathway in microseconds. Those first enzymes that bind end product have two binding sites: the active site for the substrates and an **allosteric** one to bind the end product. When the end product binds to the allosteric site, the enzyme undergoes a conformational change, inhibiting the binding of the substrates. The conformational change is reversible in the absence of the end product.

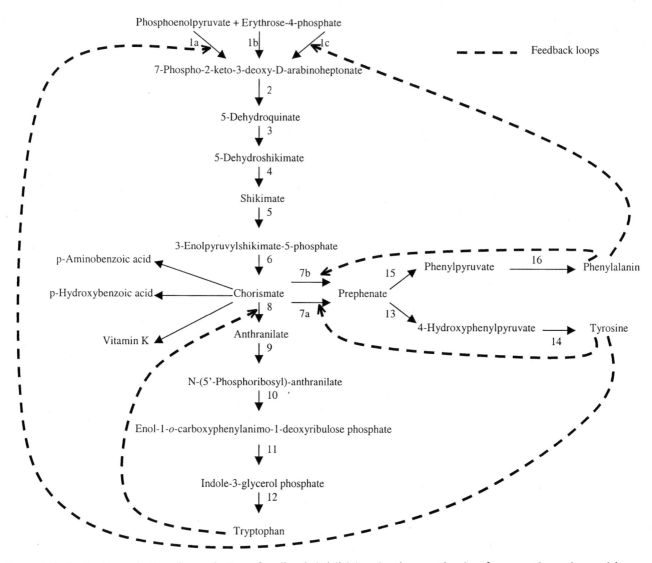

Figure 10.7. Posttranslational regulation: feedback inhibition in the synthesis of aromatic amino acids.

Topic Test 4: Posttranslational Control: Control of Enzyme Activity

True/False

1. Although synthesis of the end product of a biochemical pathway can be interrupted by enzymatic inhibition, no energy is saved in this process.

2. All the enzymes in a biochemical pathway unique to the synthesis of an end product are feedback inhibited by that end product.

Multiple Choice

3. Although end products are chemically unrelated to the substrates of a pathway regulated by feedback inhibition, an end product regulates the initial enzyme activity by
   a. binding to the active site and competing with binding of one of the substrates.

b. binding to the active site and initiating degradation of that enzyme.

c. binding to an allosteric site and initiating conformational changes in the enzyme.

d. binding to a substrate molecule changing its conformation.

e. All of the above

f. None of the above

4. All the forms of metabolic regulation are switches to channel how a cell used its energy. Which of the following phrases best expresses the type of switch that feedback inhibition (posttranslational control) represents?

a. Slowly reversible

b. Rapid but irreversible

c. Slowly becomes reversible

d. Very rapidly reversible

e. Gradually becomes constitutive

Topic Test 4: Answers

1. **False.** The energy used in the biochemical reactions to synthesize the end product is saved, although this is much less than the additional energy savings gained by not doing transcription and translation.

2. **False.** Only the first enzyme in the pathway unique to the end product is feedback inhibited by the end product.

3. **c.** Binding of the end product is to an allosteric site on the initial enzyme in the biochemical pathway and results in conformational changes inactivating the active site on the enzyme. There is no recognition between the active site and the end product.

4. **d.** Binding the end product to the initial enzyme allosterically changes the conformation of the active site. As the concentration of the end product decreases, the allosteric site on the enzyme is free, and the enzyme rapidly converts to the catalytic conformation in microseconds.

APPLICATION: SEQUENCING THE ENTIRE GENOMES OF PATHOGENS

At the end of World War II, the use of antibiotics to treat bacterial and fungal infections exploded. The belief, even among physicians and microbiologists, was that shortly bacterial and fungal infections would be an experience of the past. An antibiotic effective against any invading bacterial or fungal pathogen would be available in the physician's armamentarium. Already this hopeful statement was being challenged. Use of sulfa drugs to treat wounds during World War II was failing sometimes. Strains of bacteria initially sensitive to sulfa drugs had become resistant. In addition, some patients showed allergic responses to these drugs, necessitating changes to other antibiotics if available. The rate microorganisms have developed antibiotic resistance has increased rapidly as our use of antibiotics has increased. One strain of *Mycobacterium tuberculosis* is now resistant to all known antibiotics. However, the rate of discovery of new antibiotics has remained a lengthy haphazard process with a lot of trial-and-error experimentation.

The combination of two procedures and an understanding of protein synthesis during the 1990s opened a possible new approach to antibiotic development. First is the recognition that DNA carries a code that is transcribed to mRNA and mRNA then is translated into protein. Second is the development of experimental procedures to sequence DNA. Finally is the development of computer techniques for handling the data analysis, comparisons of DNA sequences, and three-dimensional modeling of proteins coded in the DNA sequence. Recently, an *Escherichia coli* chromosome was sequenced completely, yielding 4,638,838 DNA base pairs containing 4286 open reading frames (ORFs), 1817 of these ORFs coding for genes of known function. *E. coli* is probably the most widely used and best-known model in biologic testing. Similar total genomic sequencing is in progress (and may be complete) for *Mycoplasma pneumoniae*, agent of atypical pneumoniae; *Helicobacter pylori*, agent of stomach ulcers; *Treponema pallidum*, agent of syphilis; and *Borrelia burgdorfi*, agent of Lyme disease; and others. The hope is that by examining ORFs and determining the protein encoded, the three-dimensional structure of the protein can be determined along with its active and allosteric sites. Knowing the nature of the active or allosteric sites might improve chances of custom-designing antibiotics to interfere with the normal activities of these sites, resulting in death of the invading microorganism. These processes could reduce the cost and time to bring a useful antibiotic to market.

Chapter Test

True/False

1. At the conclusion of DNA replication, one finds a parent strand hydrogen bonded to a new progeny strand.

2. Regulation of biochemical pathways at the transcriptional level affords the greatest energy efficiency but at the cost of a slow reversal of the regulatory control.

3. In making proteins, mRNA, ribosomes, and tRNA must interact at very different times.

4. There is no start codon. The ribosome must seek the last stop codon to locate the initiating point to start protein synthesis.

5. Sigma factor must be complexed to RNA transcriptase for the recognition of individual genes or operons.

Multiple Choice

6. tRNAs enter into the translation process by
   a. positioning mRNA on the ribosome.
   b. providing the energy for translocation of a ribosome along messenger RNA.
   c. disrupting the ribosomal complex at termination of protein synthesis.
   d. bringing and positioning amino acids on the ribosomal complex for proper protein sequence.
   e. None of the above

7. How many stop codons are there?
   a. 64
   b. 61
   c. 1
   d. 3
   e. Too many to count

8. The attenuation form of gene repression in amino acid biosynthesis is activated or inhibited by
   a. the length of the leader.
   b. the size of RNA transcriptase.
   c. the concentration of that amino acid in the cytoplasm.
   d. which sigma factor recognizes the promoter.
   e. allosteric sites on RNA transcriptase.

9. Before binding to an mRNA on a ribosome, tRNA molecules must
   a. be synthesized anew.
   b. be "charged" with a specific amino acid.
   c. bind to a sugar, such as glucose.
   d. be processed in the nucleus.
   e. form polytRNA with two other tRNAs.

10. Posttranscription regulation of gene activity occurs because
    a. ribosomes bind only to monocistronic mRNA.
    b. ribosomes are unable to bridge the gaps between cistrons on polycistronic mRNA.
    c. ribosomes fail to bind to later cistrons due to improper start codons.
    d. RNases degrade mRNA molecules from terminal codons to start codons.
    e. None of the above

11. Structural genes apparently are arranged on the prokaryote chromosome as they are to provide
    a. reduced coding space on the DNA.
    b. less DNA.
    c. random assortment.
    d. coordinate control.
    e. smaller mRNA molecules.

12. The transcription of an mRNA is terminated when RNA transcriptase experiences
    a. sigma factor.
    b. elongation factor.
    c. termination factor.
    d. rho factor.
    e. All of the above

13. Bacterial DNA is replicated following which of the models given below.
    a. Rolling circle
    b. Replication bubble
    c. Reannealing after fragmentation
    d. Bidirectional
    e. Overlapping concatenates
    f. a, b, c, and e

g. c, d, and e

h. a and b

i. a and d

Short Answer

14. When are allosteric sites significant to protein function?

15. What is meant by semiconservative replication?

16. How can sigma factor be used to control gene action?

Essay

17. Discuss the role of the template in DNA or RNA replication.

18. Compare the costs and benefits of each of the three types of metabolic regulation available to prokaryotes.

Chapter Test Answers

1. **True**

2. **True**

3. **False**

4. **False**

5. **True**

6. **d**

7. **d**

8. **c**

9. **b**

10. **b**

11. **d**

12. **d**

13. **i**

14. Allosteric sites are needed to bind an inducer or a repressor to regulatory proteins that may bind to a DNA operator. In addition, allosteric sites on enzymes bind end products in feedback inhibition.

15. Semiconservative replication is observed in the replication of DNA. Each parental strand remains intact but separates as replication progresses. Each parental strand serves as a template on which a new complementary strand is synthesized. At the end of the replication, two DNA molecules have formed; each one has one parental strand and a new complementary progeny strand.

16. Gene transcription begins with the binding of sigma factor–core RNA transcriptase complex to the promoter of a gene or an operon. Different sigma factors recognize different promoters. Also, different sigma factors have greater stability under certain environmental conditions, such as heat stress, than under other conditions.

17. Templates provide a sequence of nucleotides on which a complementary strand can be faithfully replicated. Then that complementary strand serves as a template in the next round of replication to produce a complementary strand with a nucleotide sequence identical to the original template. In this manner, hereditary information is maintained generation after generation.

18. Instituting control at the transcriptional level maximizes energy savings by inhibiting mRNA and protein synthesis plus any energy used in the biochemical pathway. The cost arises in the slowness of 5 to 10 minutes to reactivate the pathway. Posttranscriptional control is faster, requiring 2 to 3 minutes to reactivate. However, this is done at the expense of transcribing mRNA. Finally, posttranslational control (feedback inhibition) has a microsecond reactivation, but energy must be expended for transcription and translation of the enzymes in the pathway.

Check Your Performance

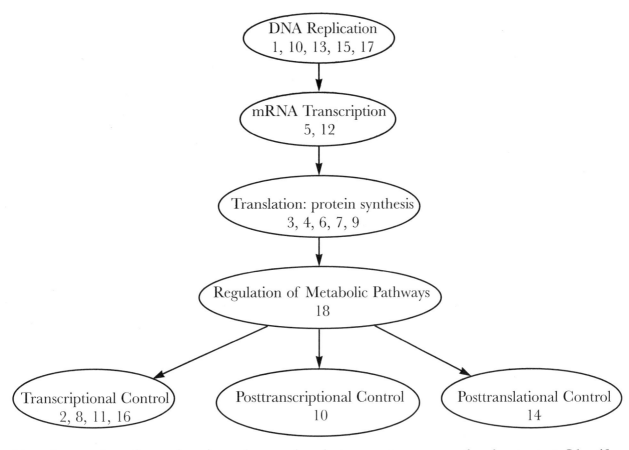

Note the number of questions in each grouping that you got wrong on the chapter test. Identify areas where you need further review and go back to relevant parts of this chapter.

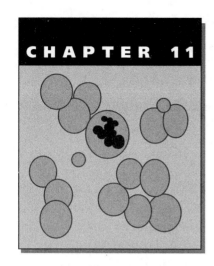

CHAPTER 11

Microbial Genetics: Transfer of Genetic Information Among Bacteria

The Darwinian-Wallace theory of evolution has as a cardinal principle that mutations occur randomly at a very low frequency. If these mutations increase an organism's survival, they will remain in the population and increase in frequency. Selective advantage means the mutant organisms are able to produce greater numbers of progeny than nonmutant organisms. These observations led to the view that organisms were conserved and underwent very slow gradual genetic and phenotypic changes.

Antibiotic resistance developed rapidly. The genes for antibiotic resistance spread through bacteria of the same species and to nonrelated species. The rapid spread of antibiotic resistance suggests that mutation is not the only genetic process involved. Genetic material is exchanged among bacteria via **plasmids**, circular DNA elements existing and replicating independently of host cell chromosomes, and **transposons**, DNA fragments containing **insertion (IS) elements** for incorporation into host cell chromosomes. Genetic alteration of bacteria occurs by three mechanisms: **transformation**, **conjugation**, and **transduction**. To incorporate genetic material from a donor into the recipient's chromosome usually requires homologous pairing of donor and recipient DNAs followed by **genetic recombination**. Some **IS elements** can insert into host chromosomes without homologous pairing.

ESSENTIAL BACKGROUND

- Bacteriophage life cycles (Chapter 9, Topic 1)
- Structure of DNA (Chapter 2, Topic 4)
- Arrangement of genes along a DNA molecule (Chapter 10, Topic 2)

TOPIC 1: PLASMIDS AND TRANSPOSABLE ELEMENTS

KEY POINTS

✓ *What characteristics are needed to be a mobile DNA fragment?*

✓ *How does each of the mobile DNA fragments differ from the others?*

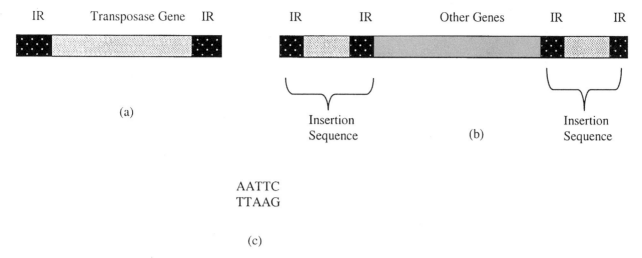

Figure 11.1a. Insertion elements (IS) bounded by insertion repeats (IR) and containing a transposase gene coding for a protein permitting chromosomal insertion. b. Transposon bounded by insertion sequences and containing genes coding for proteins unrelated to chromosomal insertion but associated with other cellular activities, such as resistance to antibiotics. c. Insertion elements and transposons target certain DNA sequences as their insertion sites, for instance transposon Tn 3 targets the sequence shown here.

IS elements are the smallest of the mobile DNA fragments, being less than 2000 base pairs (bp) long. The DNA pieces are bounded by inverse repeating DNA sequences between which there is a transposase gene coding for the protein(s) needed to introduce and remove the IS element from the host chromosome. **IS elements** target certain host DNA sequences for insertion (**Figure 11.1c**). These target sequences vary in length from 3 to 12 bp depending on which particular IS element is involved. **IS elements** contain no marker genes. Their presence must be detected by their effect on the structure of the host chromosome or the metabolism of the host cell. There are six changes IS elements induce in host chromosomes:

1. Transposition—IS element moves from one insertion site to another;
2. Excision—IS element is lost, which may reactivate an inactivated host gene;
3. Cointegration—Two IS elements join;
4. Deletion—Host DNA adjacent to IS element is lost with IS element;
5. Inversion—Host DNA adjacent to IS element is reversed in orientation;
6. Resolution—Reverse of cointegration.

Transposition results in duplication of the IS element and the target sequence. Recombinational events (cointegration, deletion, inversion, and resolution) cause no duplication of DNA sequences. When an IS element moves, it may go to another site in the host chromosome or may be exchanged between a donor and a recipient cell.

Transposons are larger mobile DNA fragments bounded by IS elements and containing other genes unrelated to transposition of the DNA element (**Figure 11.1b**). Transposons apparently arise from combinations of the six changes IS elements experience. IS elements give the transposon capacity to insert into a host cell chromosome. The "other" genes found in transposons give host organisms new metabolic capabilities, particularly to organisms in stressful environments. Some of these genes confer antibiotic resistance, carbohydrate utilization, new amino acid biosynthesis, or toxins used to overcome host resistance. Recombination or fusion of different transposons can lead to multiple antibiotic resistant genetic elements.

Plasmids have an origin of replication that allows them to exist independently of a host chromosome in the host cell's cytoplasm. Plasmids are circular like the bacterial chromosome and may contain up to 30 genes. Plasmids may exist cytoplasmically or integrate into the host cell chromosome. Some **plasmids** exist as single copies in the host cell and multiply at the same rate as the host chromosome. Other plasmids exist in multiple copies in concentrations up to 700 copies per cell. The number of copies of a plasmid in a host appears to be dependent on the nature of the plasmid and the genetics of the host cell. **Curing** occurs by slowly diluting plasmids from the growing host cell population without affecting host chromosome replication. Curing may be spontaneous or caused by environmental agents, such as acridine dyes, UV or ionizing radiation, nutrient starvation, or elevated growth temperatures. Plasmids are not genetically essential for bacteria but do improve their survival in certain environments.

Topic Test 1: Plasmids and Transposable Elements

True/False

1. Plasmids or other mobile DNA fragments can be transferred only to other cells of the same species.

2. Mobile DNA elements must pair homologously with a region on the host chromosome to insert into a host chromosome.

Multiple Choice

3. Which of the following occur(s) when an IS element inserts into a chromosome?
   a. Inversion
   b. Deletion
   c. Transposition
   d. Cointegration
   e. All of the above

Short Answer

4. Transposons differ from IS elements in what ways?

5. Of what significance are mobile DNA elements to bacterial existence?

Topic Test 1: Answers

1. **False.** Mobile DNA elements can be transferred not only between cells of the same species but also between cells of different species or even cells of different bacterial genera.

2. **False.** Some IS elements target host chromosomal sequences 3 to 4 bp long and insert using their transposase proteins. Potential insertion sites, assuming random occurrence of nucleotides, would occur every 64 to 256 bases along the chromosomal DNA.

3. **e.** All of the phenomena noted have been discovered in bacteria.

4. Transposons are larger and are bounded by complete IS elements. In addition, transposons have one or more centrally located genes that usually alter host cell metabolism by permitting host cells to use otherwise unavailable nutrients or breach the defenses of a host.

5. Bacterial cells can exist very well without mobile DNA elements. However, in certain environmental conditions, these DNA elements provide genetically determined mechanisms to overcome environmental stresses and improve bacterial survival.

TOPIC 2: GENETIC RECOMBINATION

KEY POINTS

✓ *What does genetic recombination provide bacteria?*

✓ *How is recombination achieved?*

✓ *Is there more than one mechanism for genetic recombination?*

Initially it was thought mutation, though slow, was sufficient for bacteria because the high rate of reproduction provided adequate mutant combinations for species to survive. Although mutation slowly provides changes in the genome, genetic recombination provides the testing of those changes in various genetic combinations. Several forms of genetic recombination occur in bacteria. Genetic recombination in bacteria is different from the recombination found in sexually reproducing organisms where entire genomes from two parents fuse together before recombination occurs. In bacteria only partial genomes are involved. One bacterial strain serves as a donor and a second strain is the recipient. Usually these donations are one way, not reciprocal, leading to the formation of a **merozygote**. In general, a DNA fragment, the **exogenote**, enters the recipient cell and may become a stable part of the recipient genome, the **endogenote**. A merozygote may integrate into the host chromosome, remain independent in the host cell for the life of that cell, remain independent but multiply with the host cell to form a clone of merozygous cells, or experience **host restriction** and be degraded by nucleases.

Bacteria have several forms of recombination of DNA elements: **general recombination**, a reciprocal exchange between a pair of homologous DNA elements; **site-specific recombination**, the integration of temperate viral genomes into bacterial chromosomes (Chapter 9, Topic 2); and **replicative recombination** associated with the movement of **IS elements** about the genome.

According to the Holliday model, the most accepted mechanism, **reciprocal general recombination**, requires homologously paired chromosomal regions. A recombination protein breaks one strand on each chromosome. Another recombination protein leads the broken ends to "invade" the opposing broken homologues. Sealing the newly paired broken ends forms a **chiasma** that migrates down the chromosome and rotates to form a cross-shaped structure. Breakage and sealing of DNA strands in the cross leads to recombinant DNA strands.

The Fox model is the most accepted one for **nonreciprocal general recombination**, such as observed in gene exchange in Hfr strains of bacteria (see Topic 3).

Topic Test 2: Genetic Recombination

True/False

1. Genetic recombination greatly increases genetic diversity in bacteria.

2. Homologous pairing of entire bacterial chromosomes is required before there can be genetic recombination in bacteria.

Multiple Choice

3. Reciprocal general recombination can occur
   a. any place in the bacterial chromosome.
   b. only at sites having IS elements.
   c. at very specific sites along the chromosome.
   d. only at the origin of replication.
   e. wherever a termination codon occurs.

Short Answer

4. What outcome may a merozygote experience?

Topic Test 2: Answers

1. **True.** Genetic recombination provides mechanisms for species to try new genetic combinations without waiting for chance mutation to occur.

2. **False.** Rarely is the entire bacterial chromosome involved. Usually a merozygote of less than 10% of host cell DNA is involved.

3. **a.** Homologous pairing can occur anywhere, leading to reciprocal general recombination.

4. A merozygote may become integrated into host cell chromosome, fail to integrate into the host chromosome and form a clone of merozygotic cells, or may be degraded by host cell nucleases.

TOPIC 3: CONJUGATION

KEY POINTS

✓ *What is an F plasmid?*

✓ *Under what conditions are host genes exchanged in conjugation?*

Conjugation is one of three well-characterized forms of gene exchange in bacteria and involves "fertility" genes on a plasmid. Conjugation differs from the other two forms of bacterial gene exchange in that

1. Usually many host genes are transferred from donor to recipient;

2. Cell contact between donor and recipient is necessary;

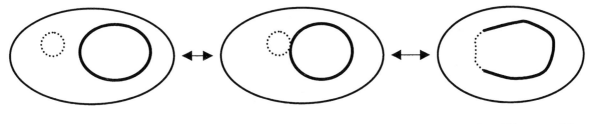

a) F plasmid independent of host cell chromosome

b) Initiation of F plasmid insertion into host cell chromosome

c) F plasmid integrated into host cell chromosome

Figure 11.2. States that F plasmid can have within a host cell. —, host chromosome; . . . , F plasmid chromosome.

3. Donor must be living;

4. Gene exchange is resistant to external DNase.

F plasmid (F for fertility) provides the genetic information for gene exchange. When F plasmid is independent, fertility genes are readily transferred, but host cell genes usually are not exchanged (**Figure 11.2**). F plasmid reversibly integrates into the chromosome at very specific sites, producing **high-frequency recombinant (Hfr)** strains. When F plasmid is integrated into the host chromosome, host genes are transferred readily. Large numbers of host genes can be exchanged between **donor** and **recipient** cells, depending on the length of time the cells are in contact. Transfer of host genes is very polarized and begins at a certain time after cell contact is achieved. Different F plasmids integrate at different specific sites on the bacterial chromosomes, leading to different Hfr strains. The order of gene transfer of the bacterial chromosome in one Hfr strain might be the reverse of gene transfer in another Hfr strain. Incorporation of F plasmid leads to several changes in the host cell, resulting in a donor cell (**F⁺** or **male** bacteria). The cell enlarges and develops several long **pili** (**pilus**, singular) that have hollow cores. These pili attach to receptor sites on recipient bacteria without the F plasmid (**F⁻** or **female** bacteria).

Topic Test 3: Conjugation

True/False

1. Exchange of chromosomal genes between donor and recipient cells in conjugation occurs in a highly polarized order.

2. F plasmids demonstrate site-directed recombination.

Multiple Choice

3. External DNase solutions fail to interrupt gene exchange in conjugation because
   a. F plasmid is coated with histones.
   b. the exchanged DNA is always within a cell wall or cell extension.
   c. DNase inhibitors are released from F⁻ cells.
   d. DNase does digest exchanged DNA, but enzymatic activity is so slow some DNA can be incorporated into an F⁻ cell nonetheless.
   e. None of the above

4. F plasmid contains the genes for
   a. DNA synthesis.
   b. glucose hydrolysis.
   c. ATP synthesis.
   d. methionine synthesis.
   e. fertility (gene exchange).

Topic Test 3: Answers

1. **True.** DNA moves from a donor cell to a recipient cell like a magnetic tape is wound from one spool on a cassette to the other. The genes resemble the electronic blips making up a song on magnetic tape.

2. **True.** Each type of F plasmid can insert itself at only one particular site on the bacterial chromosome.

3. **b.** Exchanged DNA is never free in the medium during conjugation.

4. **e.** Until F plasmid inserts into the host chromosome, it carries only the genes necessary for cell attachment and DNA exchange.

TOPIC 4: TRANSFORMATION

KEY POINTS

✓ *What is a competent cell?*

✓ *What types of genetic materials are transferred in this form of genetic exchange?*

Transformation is recognized by results from four laboratory operations:

1. Usually only one gene is exchanged at a time or if more than one is exchanged, the genes are no farther apart than 1% of the bacterial genome.

2. Cell contact is not necessary.

3. Donor does not have to be living or even present.

4. Adding DNase abolishes gene exchange.

The last observation indicates that at some point in the gene exchange, "naked" DNA is free of both donor and recipient cells. Also, if the DNA fragment has promoter and operator regions, it can be transcribed and translated in the recipient cell.

Naked DNA may arise as a result of bacterial lysis, releasing DNA fragments large enough to contain several genes (up to 10^7 daltons). Plasmids are another source of transforming DNA. Biotechnology using recombinant DNA is based on introducing artificially constructed plasmids into suitable host cells (see Chapter 20, Topic 1). For naked DNA to be accepted into a host cell, the host cell must be **competent**. In naturally competent cells a small protein, competence factor, activates synthesis of 8 to 10 proteins needed for transformation. Transformation is found scattered in some gram-positive and gram-negative bacteria. Competent cells have cell surface changes, in particular a DNA-binding protein, a "nickase," and a competence-specific protein (**Figure 11.3**). Metabolic energy is required to incorporate DNA into the competent cell. Com-

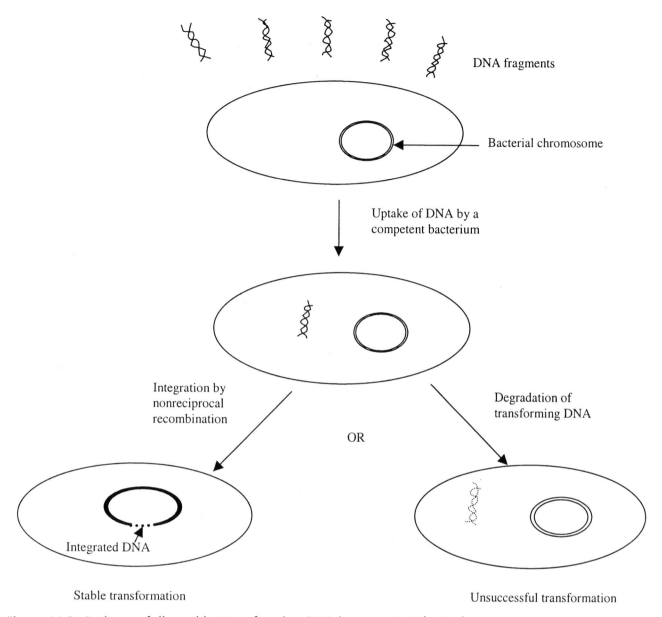

DNA fragments

Bacterial chromosome

Uptake of DNA by a
competent bacterium

Integration by
nonreciprocal
recombination

Degradation of
transforming DNA

OR

Integrated DNA

Stable transformation

Unsuccessful transformation

Figure 11.3. Pathways followed by transforming DNA in competent bacteria.

petence fluctuates during the cell cycle. In flask cultures, competent cells are most numerous during mid-log to early stationary phases. Cells not naturally competent can be induced artificially to accept DNA fragments or plasmids. When using *Escherichia coli*, heat shocking or electroporation can induce cells to incorporate DNA fragments. Coupling transformation with artificial construction of plasmids with desirable traits forms the core of the biotechnology revolution (see Chapter 20).

Topic Test 4: Transformation

True/False

1. Only naturally competent cells can be transformed.

2. Naturally transformable cells are found widely among bacteria.

Multiple Choice

3. Transformation can be blocked by the addition of what to the medium?
   a. α-Amylase
   b. Proteinase K
   c. Deoxyribonuclease
   d. Ribonuclease
   e. Single-stranded binding protein

4. Which of the following do naturally competent cells *not* need?
   a. β-Lactamase
   b. Competence factor
   c. "Nickase"
   d. DNA-binding protein
   e. Competence-specific factor

Short Answer

5. Why is cell contact or a living donor not necessary for transformation?

Topic Test 4: Answers

1. **False.** By altering the medium of the bacteria and/or using "leaky" membrane mutants, one can obtain transformation in bacteria not known to be competent.

2. **False.** Although bacterial species demonstrating natural competence are scattered through the Eubacteria, they are restricted to certain genera.

3. **c.** Deoxyribonuclease (DNase) can degrade the cell-free DNA fragments and plasmids occurring in transformation.

4. **a.** β-Lactamase is an enzyme that degrades penicillin and is produced by certain penicillin-resistant bacteria.

5. Because in transformation the DNA elements being exchanged are cell free, the source of the DNA (the donor) need not be present.

TOPIC 5: TRANSDUCTION

KEY POINTS

✓ *How does host DNA become incorporated into a virus coat?*

✓ *How can the two types of transduction be differentiated?*

Transduction, the third form of bacterial gene exchange, is recognized by the following four laboratory procedures:

1. Usually only one gene is exchanged at a time or if more than one gene is exchanged, the genes must be within 1% of the bacterial genome length.

2. Cell contact is not necessary.

3. Donor cell need not be living.

4. DNase does not block gene exchange.

Because cell contact is not required and the donor cell need not be living, exchangeable DNA must be free of host cells for a period during transduction. The size of DNA exchanged and its protection from DNase suggests that a virus or a protein coat of a virus is involved. Two forms of transduction exist, each arising from mistakes in viral activity in a host cell: **generalized transduction** and **specialized transduction**.

In **generalized transduction**, any bacterial gene in the genome may be exchanged. Some viruses do not activate a DNase to degrade the host chromosome when they invade a bacterium (Chapter 9, Topic 1). These viruses replicate their DNA in chains of genomes, **concatenates**. These concatenates resemble the host chromosome. When packaging and maturational proteins assemble progeny viruses, they cut long DNA into virus genome-long pieces. Distinction between virus and chromosomal DNA is not made, and both DNA sources become incorporated into virus coats (**Figure 11.4**). Any DNA fragment with promoter and operator regions can be tran-

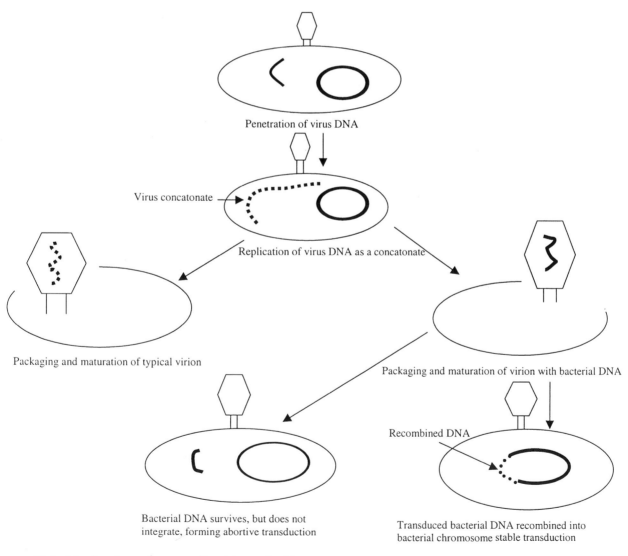

Figure 11.4. Mechanism of generalized transduction.

scribed and translated. Transduced DNA very rarely has an origin of replication. Therefore, if the transduced DNA fragment is not recombined into the host chromosome, it will not be replicated and only one copy of the transduced gene will exist in a bacterial colony. Small **abortive** colonies may arise because of cross-feeding of a necessary nutrient by the cell with the transduced DNA. On the other hand, if the transduced DNA is incorporated into the host chromosome, it will be replicated as the host chromosome replicates. A large colony of cells, each with the transduced DNA, results leading to **complete** transduction.

In **specialized transduction**, only certain host genes are exchanged. Temperate bacteriophages may be lytic or may become lysogenic and insert their DNA by site-specific recombination into the host chromosome (Chapter 9, Topic 1). The **prophage** can exist for variable periods of time in the host chromosome. However, excision of the prophage can occur spontaneously or in response to environmental stress. When illegitimate crossing over occurs, precise excision of the prophage fails. This produces a virus genome containing host chromosome from either side of the insertion point (**Figure 11.5**). This degenerate phage will replicate, and many copies will be enclosed in viral protein coats. Hybrid viruses can become lysogenic on penetration of uninfected host cells. They insert exchanged bacterial genes into the host chromosome. This mechanism limits gene exchange only to those genes on either side of the prophage.

Topic Test 5: Transduction

True/False

1. Any bacterial gene can be exchanged in generalized transduction.

2. The size of the virus coat limits the amount of bacterial DNA that can be exchanged during generalized transduction.

Multiple Choice

3. The DNA exchanged in transduction is resistant to DNase degradation because
   a. it stays within the donor cell.
   b. it passes through a bridge between donor and recipient cells.
   c. it passes along the hollow core of a pilus connecting donor and recipient cells.
   d. it is enclosed in a virus coat.
   e. None of the above

4. What event leads to the deposition of bacterial DNA in a virus coat in generalized transduction?
   a. Failure of a temperate bacteriophage to enter the lysogenic state
   b. Mistaken recognition of bacterial chromosome as a viral concatenate and its packaging into virus coats
   c. Illegitimate crossing over in excision of the prophage
   d. Inability of the temperate bacteriophage to attach to a host cell
   e. Only partial penetration of a host cell by the temperate bacteriophage

Short Answer

5. Why is specialized transduction restricted to exchange of only certain bacterial genes?

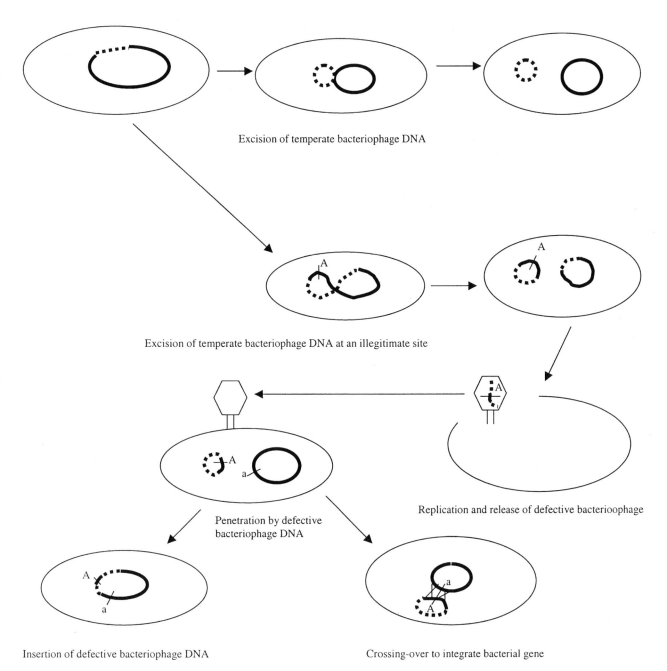

Excision of temperate bacteriophage DNA

Excision of temperate bacteriophage DNA at an illegitimate site

Replication and release of defective bacterioophage

Penetration by defective
bacteriophage DNA

Insertion of defective bacteriophage DNA

Crossing-over to integrate bacterial gene

Figure 11.5. Mechanism of specialized transduction.

Topic Test 5: Answers

1. **True.** Because viral maturational and packaging enzymes fragment long DNA, different fragments of bacterial DNA may be incorporated into viral coats.

2. **True.** The size of the viral coat determines the amount of DNA it can hold.

3. **d.** The lack of cell contact and need for a living donor cell indicated the DNA exchanged must be extracellular for part of the exchange. Protection from DNA degradation indicates the DNA is covered.

4. **b.** Bacterial chromosomes and viral concatenates are not distinguished by the maturation and packaging enzymes of temperate bacteriophages.

5. Because the hybridization of viral and bacterial DNAs occurs as a result of illegitimate crossing over, the only components of the bacterial genome that could be included in the viral genome would be the regions adjacent to either side of the prophage insertion site.

APPLICATION: THE ANTIBIOTIC PANACEA CLOUDED BY RESISTANCE TRANSFER FACTORS

The end of World War II permitted the launching of the use of antibiotic therapy to control pathogenic organisms of several serious human diseases, such as bacterial pneumonia and tuberculosis. Soon use of antibiotics became widespread, being used in treatment and prophylaxis for many types of diseases for one reason or another. For instance, penicillin was used by persons with head colds (rhinovirus infections) for which penicillin is ineffective. Penicillin was used very effectively in many parts of the world for treating intestinal infections where water is contaminated with human sewage. However, in the 1950s, Japanese microbiologists began encountering strains of *Shigella*, the agent of bacterial dysentery, having resistance to individual antibiotics. Shortly, they found *Shigella* strains resistant not only to individual antibiotics but also to several. Close examination demonstrated the presence of R plasmids. Even scarier is the exchange of R plasmids from fairly rare pathogens to more common and widespread genera, such as *Escherichia* and *Salmonella*.

Meanwhile, antibiotics have come to be used as a panacea, even for nonhuman prophylactic applications in agriculture and aquaculture. The result has been a selective process favoring antibiotic-resistant microbes and stimulating exchange of R plasmids. Today the presence of resistant microbes is near ubiquitous. Some inexpensive formerly effective antibiotics are essentially useless today. The warning is "take heed" in the use of antibiotics. Microorganisms appear to gain resistance faster than humans can develop new antibiotics.

Chapter Test
True/False

1. All genetic elements that are exchanged between bacteria must have origins of replication.

2. In most instances the size of the DNA element transferred is less than 1% of the bacterial genome.

3. Any DNA fragment having promoter and operator regions can be transcribed and translated in the recipient cell cytoplasm.

4. Exogenotes must be incorporated as endogenotes in bacterial chromosomes before genetic information is used.

Multiple Choice

5. Which form of genetic exchange among microorganisms is *not* found among bacteria?
   a. Parasexual exchange
   b. Transformation
   c. Conjugation
   d. Transduction
   e. None of the above

6. Which of the following types of chromosomal changes is *not* induced by IS elements?
   a. Inversion
   b. Cointegration
   c. Duplication
   d. Transposition
   e. Deletion

7. The mobility between cells shown by IS elements and transposons resembles which of the recognized forms of bacterial gene exchange?
   a. Conjugation
   b. Transformation
   c. Duplication
   d. Transduction
   e. Chromosomal fragmentation

8. Plasmids are able to do which of the following that cannot be accomplished by other mobile genetic elements?
   a. Insert into a chromosome
   b. Self-replicate
   c. Transfer antibiotic resistance
   d. Experience recombination
   e. None of the above

9. Strain that donates bacterial genes at a high rate is a(an) _____ strain.
   a. Female (F⁻)
   b. RTF
   c. Male (F⁺)
   d. Hfr
   e. Col

10. Which of the following induces competence in normally noncompetent cells?
    a. Glucose
    b. Heat shock
    c. Starvation
    d. Virus invasion
    e. All of the above

Short Answer

11. Which form of bacterial gene exchange probably is most beneficial to a particular species?

12. How do multiantibiotic resistant R plasmids probably arise?

Essay

13. What is gained by permitting gene exchange in species like bacteria that have high rates of replication?

Chapter Test Answers

1. **False**
2. **True**
3. **True**
4. **False**
5. **a**
6. **c**
7. **b**
8. **b**
9. **d**
10. **b**
11. Conjugation because many bacterial genes may be exchanged in a very organized fashion from a single event.
12. By the cointegration of transposons and/or plasmids carrying different antibiotic-resistance genes.
13. Although bacteria multiply at a terrific rate and hence produce many mutants, advantageous mutants most likely would occur in different cell lines. To gain multiple antibiotic resistance to five antibiotics, for example, would require five very rare mutations in the same cell line. Even with the rapid multiplication of bacteria, considerable time would be required to accumulate five such mutations. On the other hand, gene exchange permits a more rapid accumulation of genes, bringing together genetic information from several cell lines. Thus, bacteria gain information from across generic or familial taxonomic lines, providing an enormous pool of variation for selection to influence.

Check Your Performance

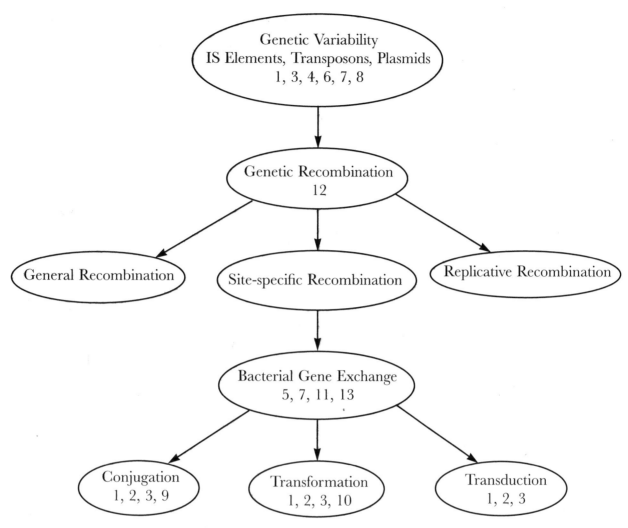

Note the number of questions in each grouping that you got wrong on the chapter test. Identify areas where you need further review and go back to relevant parts of this chapter.

Midterm Exam

Chapter 1–5 Multiple Choice

1. Identification of causative agents of infectious disease was made possible by application of a set of criteria developed by
   (a) Robert Hooke.
   (b) Robert Koch.
   (c) Antoine Leeuwenhoek.
   (d) Louis Pasteur.

2. The volume of a typical eukaryotic cell is _____ times the volume of a typical prokaryotic cell.
   (a) 10
   (b) 100
   (c) 1000
   (d) 10,000

3. Chemicals that readily undergo oxidation
   (a) have relatively low electronegativities.
   (b) have a high tendency to lose electrons.
   (c) are good terminal external electron acceptors in respiration.
   (d) have relatively high reduction potentials.

4. The formation of a _____ ion from a neutral atom occurs when it is oxidized.
   (a) positive
   (b) negative

5. Teichoic acid is associated with the surface of
   (a) Gram negative bacteria.
   (b) Gram positive bacteria.
   (c) prokaryotes in the Archaea domain.
   (d) all of the above.

6. Which of the following bacterial components does NOT function partly in adhesion to surfaces, and/or to other bacteria?
   (a) capsules
   (b) fimbriae
   (c) flagella
   (d) pili

7. Plasmolysis
   (a) describes water movement out of a cell.
   (b) is a consequence of active transport.
   (c) does not occur unless there is an energy supply.
   (d) results in the swelling of a cell and possible lysis.

8. Turbidity is a property of a liquid suspension of bacteria that can be used to measure the
   (a) number of viable bacteria.
   (b) bacterial mass.
   (c) both of these.

9. Compatible solutes accumulate in the cytoplasm
    (a) for storing carbon, energy, and/or phosphate.
    (b) in response to increases in external osmotic pressure.
    (c) to protect bacteria from swelling and bursting in a hypotonic medium.
    (d) to increase the buoyancy of the cell.

10. A bacterium that is growing lithotrophically and autotrophically uses _____ compounds as its external source of electrons, and _____ compounds as its external source of carbon.
    (a) organic :: organic
    (b) organic :: inorganic
    (c) inorganic :: organic
    (d) inorganic :: inorganic

11. Fermentation
    (a) cannot occur in the presence of oxygen.
    (b) uses either an external or an internal electron acceptor.
    (c) uses ATP synthase for phosphorylating ADP to form ATP.
    (d) is none of the above.

12. Oxygenic photosynthesis
    (a) requires oxygen to function.
    (b) results in an increase in atmospheric carbon dioxide.
    (c) uses ATP synthase for phosphorylating ADP to form ATP.
    (d) uses an organic compound as an external source of electrons.

True/False

1. Prokaryotes are only found in the domain Bacteria.

2. When a protein has quaternary structure, it consists of two or more polypeptides.

3. ATP is the energy source for flagellar rotation.

4. Most bacteria are pathogenic.

5. Photosynthetic organisms primarily use substrate-level phosphorylation for the formation of ATP.

6. Bacteria grow asynchronously in a chemostat.

7. It is incorrect to say that an object is "almost sterile."

Short Answer

1. Identify the domain of organisms most commonly found growing in extreme environments of temperature, salinity, and pH.

2. List two parameters that influence the Gibb's free energy change associated with the coupling of two half reactions in a redox reaction.

3. What property of lipids makes them ideal structural components of membranes?

4. Name two highly reactive free radicals arising from oxygen metabolism in cells.

5. _____ solutes are used by bacteria to increase the osmotic pressure of their cytoplasm.

6. List five important features of the chemiosmotic theory explaining oxidative phosphorylation.

Chapter 6

1. Examples of amphibolic pathways are
   a. Glycolysis.
   b. Tricarboxylic Acid Cycle.
   c. Pentose Phosphate Pathway.
   d. a and b.

2. The Entner-Duodoroff pathway occurs in
   a. procaryotes.
   b. eucaryotes.
   c. fungi.
   d. plants.

Chapter 7

1. Classification of bacteria is based on
   a. physiological characteristics.
   b. cultural characteristics.
   c. biochemical characteristics.
   d. morphological characteristics.
   e. all of the above.

2. Binary data in bacterial classification is
   a. data that has multiple possible results, such as the size of the zone of inhibition.
   b. data on two different characteristics.
   c. data that has only two possible results, such as Gram + and Gram −.
   d. none of the above.

Chapter 8

1. Bacteriostatic means
   a. bacterial static charges.
   b. bacterial growth is inhibited.
   c. bacterial death.
   d. it is produced by bacteria.

2. Examples of metabolic analogues are
   a. sulfa drugs.
   b. AZT.
   c. penicillin.
   d. a and b.

Chapter 9

1. Viruses differ from other microbes in
   a. having only DNA or RNA.
   b. having independent metabolism.
   c. having subcellular organelles.
   d. having cytoplasm.

2. Bacterial DNA with integrated prophage is called a
   a. coliphage.
   b. temperate phage.
   c. lysogen.
   d. page phage.

Chapters 10–11 True/False

1. A codon consists of two amino acids. There are 16 codons to code for the nucleotides in a protein.

Multiple Choice

2. What form of regulation would be most energy efficient for an organism to use to regulate degradation of a carbohydrate?
   a. Gene repression
   b. Feedback inhibition
   c. Catabolyte
   d. Gene induction
   e. Translational control

3. An operon consists of which of the following regions of DNA?
   a. Structural genes
   b. Regulator gene
   c. Sigma factor
   d. Operator
   e. Promoter
   f. All of the above
   g. a, d, and e
   h. b, c, and d
   i. a, c, and d

Short Answer

4. Why is feedback inhibition not as energetically efficient as transcriptional controls?

Midterm Exam Answers

Chapters 1–5 Multiple Choice Answers

1. **b**   2. **c**   3. **c**   4. **a**   5. **b**   6. **c**   7. **a**   8. **b**   9. **b**   10. **d**   11. **d**   12. **c**

Chapters 1–5 True/False Answers

1. **F**   2. **T**   3. **F**   4. **F**   5. **F**   6. **T**   7. **T**

Chapters 1–5 Answers to Short Answer Questions

1. The Archaea domain

2. (1) the difference in reduction potentials of the two half reactions.
   (2) the number of electrons transferred between the two half reactions.

3. They have both hydrophobic and hydrophilic regions, and thus provide an interior hydrophobic region of the membrane with hydrophilic regions on the internal and external surfaces of the membrane.

4. superoxide radical and hydroxyl radical

5. compatible

6. (1) the membrane is impermeable to protons
   (2) a respiration chain that provides oxidation-reduction driven translocation of protons across the membrane
   (3) an external electron acceptor
   (4) a proton gradient that forms an electrochemical proton motive force
   (5) an ATP synthase imbedded in the membrane using energy from the proton motive force for oxidative phosphorylation

Chapter 6

1. **d**   2. **a**

Chapter 7

1. **e**   2. **c**

Chapter 8

1. **b**   2. **d**

Chapter 9

1. **a**   2. **c**

1. **F**   2. **d**   3. **g**

4. An organism will synthesize the mRNAs and enzymes needed in a metabolic pathway if feedback inhibition is used for regulation, whereas these molecules will not be synthesized when the pathway is under transcriptional controls.

UNIT III

MEDICAL MICROBIOLOGY, IMMUNOLOGY, ENVIRONMENTAL AND APPLIED MICROBIOLOGY

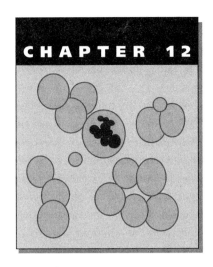

CHAPTER 12

Symbiotic Associations, Microbial Pathogenicity, and Epidemiology

Organisms relate to each other in a variety of ways. The relationship may benefit both populations, benefit one and have no effect on the other, or benefit one and harm the other. When one organism or population is harmed, disease is the result.

Obviously, to study how populations of organisms relate to each other, the organisms *cannot* be studied in pure culture. In the lab, chemostats may contain mixed culture of bacteria in an attempt to simulate the real environment. Other studies of the interaction between organisms are done by analysis of what happens outside the lab. Because of the complexity of environments, this does not give a true picture of the relationships between populations. Other studies of the interactions between organisms include the virulence factors that allow pathogens to produce disease in their hosts.

ESSENTIAL BACKGROUND

- General knowledge of microbiology
- Knowledge of structure of bacterial cells
- General knowledge of animal anatomy

TOPIC 1: COMMENSALISM, MUTUALISM, AND PARASITISM

KEY POINTS

✓ *What is commensalism?*

✓ *What is mutualism?*

✓ *What is parasitism?*

✓ *What are some typical examples of each kind of relationship?*

The term "**commensalism**" comes from the Latin words *cum* (combining form "com"), meaning "with," and *mensa*, meaning "table." Organisms that have a commensal relationship are described as organisms that are "eating at the same table."

In commensal relationships, one organism or population is neither harmed nor helped by the relationship. The other organism or population gains. The relationships generally are not obligatory relationships. The beneficiary could form a symbiotic relationship with another benefactor population that fulfilled the same needs for the beneficiary.

Mutualism is a relationship in which both populations of organisms benefit from the association. In many cases, mutualism allows two populations of organisms to live in an environment that neither could live in alone. In mutualistic relationships, the two populations of organisms may be said to be acting as one.

Chemotaxis is one factor that brings organisms together in mutualistic relationships. Recall that chemotaxis occurs when organisms move toward or away from each other in response to the chemicals around the other organism. The environmental chemicals form a gradient, with the highest concentration immediately surrounding the organisms that excrete the chemical.

Parasitism is a relationship that benefits one population at the expense of the other. Parasitic relationships are often extremely **specific**. Virus/host relationships are an example of this specificity. Most viruses infect only a small range of host species. For more information on viruses, see Chapters 9 and 17.

In parasitic relationships, the parasite gets a suitable environment for growth. The host provides for the parasite's nutritional needs either as "food" itself or the parasite uses the host's food and/or metabolic products from the host. In the extreme case of viruses, the host even provides the mechanism for the parasite to reproduce itself.

The parasite may also be a host in a relationship called **hyperparasitism**. An example of hyperparasitism is *Bdellovibrio* species that parasitizes some gram-negative bacteria. *Bdellovibrio* species can be parasitized by bacteriophage.

When an organism is engulfed by another, it is called **predation**. In this case, rather than a host–parasite relationship, it is a predator–prey relationship. An example of predation is when protozoans engulf algae and bacteria as food. **Table 12.1** summarizes symbiotic relationships.

Topic Test 1: Commensalism, Mutualism, and Parasitism

True/False

1. Syntrophism is a type of mutualism.

2. Synergism is an obligate relationship.

Multiple Choice

3. Commensalism means
   a. growing together.
   b. eating at the same table.
   c. common metabolism needs.

4. A commensal relationship may exist because
   a. both populations have the same nutritional needs.
   b. both populations produce nutrients the other population needs.
   c. one population excretes a growth factor needed by the other population.

	POPULATION 2 OR ORGANISM 1		POPULATION 2 OR ORGANISM 2	
RELATIONSHIP	HARMED/HELPED/ NEUTRAL	GAINS FROM RELATIONSHIP	HARMED/HELPED/ NEUTRAL	MAY GAIN FROM RELATIONSHIP
Commensal	Neutral	Nothing	Helped	Metabolites Vitamins Growth factors Anaerobic environment Freed nutrients from environment Compounds converted to utilizable form Toxins removed from environment
Dometabolism	Neutral	Nothing	Helped	Organism 1 oxidizes a substrate that it cannot use, Organism 2 needs the substrate
Mutualism				
Organisms often brought together by chemotaxis				
Obligate	Helped	Nutrients Needed environment Toxins removed	Helped	Nutrients Needed environment Toxins removed
Synergism	Helped	Same as above	Helped	Same as above
Syntrophism— may be obligate	Helped	Same as above	Helped	Same as above
Endosymbiosis— one population lives inside the other	Helped	Same as above	Helped	Same as above
Parasitism				
Extremely specific	Harmed	Nothing	Helped	Nutrients, environment

Table 12.1. Symbiosis

5. Mutualism is a relationship in which
   a. both populations benefit from the relationship.
   b. neither population benefits from the relationship.
   c. one population benefits from the relationship and the other population is harmed by the relationship.
   d. one population benefits from the relationship and the other population is neither harmed nor benefited by the relationship.

6. Hyperparasitism is
   a. an organism being extremely parasitic to its host.
   b. an organism that engulfs other organisms.
   c. an organism being a parasite of one organism and the host to a different parasitic organism.

Short Answer

7. Define parasitism.

Topic Test 1: Answers

1. **True.** Syntrophism is a type of mutualism.

2. **False.** Synergism is not an obligate relationship.

3. **b**

4. **c**

5. **a**

6. **c**

7. Parasitism is a relationship in which one organism population benefits and the other is harmed.

TOPIC 2: TYPES OF PATHOGENS

KEY POINTS

✓ *How do pathogens differ from opportunist pathogens?*

✓ *What circumstances give opportunist pathogens the opportunity to be pathogenic?*

A good way to think about the relationship between pathogenic organisms and their hosts is to consider it an ever-escalating war. The microbe invades the host. The host's phagocytic cells attack the invaders. The invading microbe develops enzymes or toxins to kill the phagocytic cells. The host develops antibodies against the invading microbes. The invading microbes develop enzymes that destroy the antibodies. Modern medicine finds antibiotics that will kill the invading microbes. The invading microbes develop resistance to the antibiotics. And so the war continues.

Microorganisms can be divided into two categories: **pathogens** and **nonpathogens**. The pathogen category can be further subdivided into pathogens and **opportunist pathogens**.

Organisms classified as pathogens are always associated with disease. Their presence in the infected host is considered an abnormal state. Pathogens are parasitic. They benefit from the relationship and harm the host. Examples of these organisms are *Mycobacterium tuberculosis* and *Yersinia pestis*.

Opportunist pathogens are part of the normal flora and become pathogenic only when they have the **opportunity** to be pathogenic, hence their classification opportunist pathogens. Examples of opportunist pathogens are *Escherichia coli*, *Pseudomonas species*, some molds, and yeasts.

Opportunist pathogens are normally commensal or mutualistic organisms, but they can be pathogenic if the opportunists get into locations in the body where they are not normally found, such as *E. coli* in a wound, or opportunists infect debilitated or immune suppressed hosts.

Finally, there are nonpathogenic organisms. These organisms are, as the classification suggests, organisms that never produce disease. These organisms have a commensal or mutualistic relationship with their host or benefactor.

Topic Test 2: Types of Pathogens

True/False

1. Opportunist pathogens are part of the host's normal flora.

2. Organisms that are classified as pathogens are part of the host's normal flora.

3. One way opportunists may cause disease is when the host is debilitated.

Multiple Choice

4. Organisms whose presence is always abnormal are clasified as
   a. nonpathogens.
   b. opportunist pathogens.
   c. pathogens.

5. Organisms that never cause disease are called
   a. nonpathogens.
   b. opportunist pathogens.
   c. pathogens.

Topic Test 2: Answers

1. **True**

2. **False.** Organisms that are classified as pathogens are not part of the host's normal flora.

3. **True**

4. **c**

5. **a**

TOPIC 3: PROPERTIES THAT DETERMINE THE DEGREE OF VIRULENCE OF AN ORGANISM

KEY POINTS

✓ *What are the two general types of toxins produced by bacteria?*

✓ *How do enzymes produced by bacteria allow them to invade the host tissue?*

Epithelial tissue is **skin** or **mucous membranes** that line the natural body orifices. Generally, the epithelial tissue covering the body and lining the natural cavities prevents parasitic pathogenic organisms from attacking. The fluids that constantly wash the mucous membranes wash away microorganisms. Bathing, hand-washing, clothes rubbing against skin, even walking outside on a windy day remove organisms from skin as skin cells are lost.

Pathogens use a variety of methods for staying **attached** to and **invading** skin and mucous membranes. The methods of attachment are categorized as adherence factors. Organisms that are not able to invade host tissue are not pathogenic. Being able to attach is the first and most important factor in whether or not an organism causes disease. If the organisms cannot attach, they cannot invade.

The site of invasion is usually epithelial tissue. In the case of surgical or deep wounds, other tissue may be the site of invasion. Pathogens use several **adherence factors** to stay attached to the epithelial tissue. Some of the common adherence factors are hydrophobicity, net surface charge (remember opposite charges attract), ligands, host cell receptors, bacterial cell structures, and external polysaccharides, glycocalyx, capsules, and slime layers.

Other factors, referred to as **virulence factors**, allow pathogens to cause disease. Many organisms have environmental "triggers" that activate the virulence factors. The environmental triggers include temperature, pH, the pathogen's growth phase, and the availability of specific nutrients.

Table 12.2. Differences Between Endotoxins and Exotoxins

Toxin Type	Source	Type of Molecule	Heat Stability	Antibody Production	Gram Reaction	Immunization Available	Organisms that Produce Exotoxins**
Exotoxin	Excreted by organism	Protein	Destoyed by heat	Good	G–/G+	Yes*	*Corynebacterium diphtheriae* *Clostridium* species *Staphylococcus aureus* *Streptococcus pyogenes* *Vibrio cholera* *Yersinia enterocoliticus* *Vibrio parahemyliticus* *Aeromonas* species
Endotoxin	Part of cell wall	Lipopolysaccharide	Stable	Poor	G– only	None	

* Examples of immunization against exotoxins are the vaccines for tetanus and diphtheria.
** Exotoxins often act as enterotoxins because they cause vomiting and diarrhea.

Table 12.3. Enzymes and Their Targets

Enzyme	Attacks	Target Substance Location	Category of Action
Lecithinase	Lecithin	Host cell membranes	Spread of infection
Collagenase	Collagen	Host's connective tissue	Spread of infection
Hyaluronidase	Hyaluronic acid	Cement between cells	Spread of infection
Streptokinase (also called fibrinolysin)	Fibrin	Blood clots	Spread of infection
Cytolysin	Cells	Lyses cells	Spread of infection and protection against immune system
DNase	DNA	Host DNA	Spread of infection
IgA1 proteases	Protein	Type 1 IgA secretory antibody	Protects from host defenses
Coagulase	Blood	Clots blood to wall off bacteria from host defenses	Protects from host defenses

Invasiveness is an extremely important virulence factor. The greater the pathogen's ability to invade host cells and tissues, the greater its chance of producing disease. Bacteria produce enzymes and toxins that facilitate invasion of the host.

Two categories of toxins are produced by bacteria: exotoxins and endotoxins. The differences between endotoxins and exotoxins are summarized in **Table 12.2**.

Enzymes are another virulence factor or tool that bacteria use for invading the host tissue. Enzyme action and targets are shown in **Table 12.3**.

Other virulence factors include antiphagocytic factors, intracellular pathogenicity, and antigenic heterogenicity.

Topic Test 3: Properties that Determine the Degree of Virulence of an Organism

True/False

1. Bacteria produce enzymes that attack host cells or specific structures of the host cells.

2. Endotoxins are polypeptides.

3. Enterotoxins are exotoxins.

Multiple Choice

4. Endotoxins are
   a. part of gram+ cell wall.
   b. part of gram− cell wall.
   c. proteins.

5. Exotoxins are
   a. part of bacterial cell walls.
   b. toxins excreted by bacteria.
   c. leukocytes.
   d. proteins.
   e. b and d

Short Answer

6. What type of bacterial toxin is excreted by the organism?

Topic Test 3: Answers

1. **True.** Bacteria produce enzymes that attack host cells or specific structures of the host cells.

2. **False.** Endotoxins are lipopolysaccharides.

3. **True.** Enterotoxins are exotoxins.

4. **b**

5. **e**

6. An exotoxin.

TOPIC 4: SOURCE AND ROUTE OF SPREADING OF MICROBIAL PATHOGENS

KEY POINTS

✓ *How do organisms spread from host to host?*

✓ *How do organisms enter the host?*

✓ *How can organisms be contained?*

Diseases are described as having two **directions of spread: horizontal** transmission, or person to person, and **vertical** transmission, or parent to offspring. **Reservoirs** of infection include animals; soil, water, and food; and humans. **Carriers** are animals and humans who spread the disease but have no symptoms of it. The three categories of a carrier are a healthy carrier, someone with no symptoms; a chronic carrier, someone who had the disease and spreads it, long after recovering; and an intermittent carrier, someone who sheds organisms intermittently.

Zoonosis is a disease that is transmitted from an animal to a human.

Vector is a carrier of a disease. Insects are common vectors. They can carry the disease producing organisms on their bodies or be part of the life cycle of the disease producing organisms.

Nosocomial infections are infections a person gets in the hospital or other type of medical facility. The two categories of nosocomial infections are exogenous, such as an infection from droplets, direct contact with infected body fluids, bandages, or nonsterile medical equipment; and endogenous, when an opportunist pathogen in the patient's body takes the opportunity to cause an infection.

There are portals of entry and exit for the infection and the infection's spread through the body. The body is covered by epithelial tissue—skin and mucous membranes. Once the pathogens reach the host, they invade the host through any opening in the host's body. There are two general types of body openings: natural and artificial. Natural openings or orifices include the mouth, nose, ear canal, tear ducts, urethra, vagina, seminal vesicles, milk ducts, and anus. The natural openings allow the invading organisms access to the mucous membranes and to the internal organ systems that open directly to the outside. The natural openings and exposed mucous membranes are all bathed with body fluids, such as mucus, tears, saliva, urine, semen, or gastric enzymes. Artificial body openings are the holes created in the skin or mucous membranes by some object, such as an insect's stinger, teeth, or surgical instruments. The artificial body openings always open the skin or mucous membranes to invading organisms. They may also open the subcutaneous tissues and organ systems to the invading microbes or allow the microbes access to the bloodstream. Once organisms get into the bloodstream, there is the chance they will cross the blood–brain barrier and get into the cerebrospinal fluid.

The pathogens spread through the host by attacking the tissues or by spreading through the lymphatic system and bloodstream. Tissue infections may be localized at the site where the organisms entered the host's body. Infections spread through the blood and lymphatic systems are generalized.

When pathogens are found in the blood, they may be multiplying in the blood, which is septicemia. Alternatively, the blood may only be transporting the bacteria to a site where they can start an infection, a bacteremia condition. Eventually, the invading pathogens leave one host and move to another host. The portals of exit for microorganisms are the same as the portals of entry.

Infections can be controlled by several methods: containment by isolation and quarantine, immunization, and vector control.

Topic Test 4: Source and Route of Spreading of Microbial Pathogens

True/False

1. A disease that the fetus gets *in utero* is an example of vertical transmission.

Multiple Choice

2. An infection that a patient acquires from nonsterile surgical instruments during surgery is a

a. zoonosis.

b. fomite infection.

c. nosocomial infection.

Short Answer

3. How does a bacteremia differ from a septicemia?

4. How does a zoonosis differ from a vector?

Topic 4 Test: Answers

True/False

1. **True**

2. **c**

3. In a septicemia, the bacteria are growing in the bloodstream. In a bacteremia, the blood transports the bacteria to a different site where they will set up an infection.

4. A zoonosis is an infection that a person gets from a sick animal. A vector is an insect carrying the bacteria and depositing the organisms on food by landing on the food or when an insect is part of the pathogen's life cycle.

TOPIC 5: EPIDEMIOLOGY: CAUSES AND DISTRIBUTIONS OF INFECTIOUS DISEASES

KEY POINTS

✓ *What is epidemiology?*

✓ *What do pandemic, epidemic, and endemic mean?*

Epidemiology is the study of the factors that affect **disease statistics**. The epidemiologist looks at the factors involved in the occurrence and spread of disease within populations of humans and animals.

A **communicable disease** is one that can be spread before the infected person has any symptoms of the disease and during the recovery period. People who spread the disease are carriers or hosts to the pathogen.

Herd immunity is when most of the members of a population are immune to a disease. Herd immunity is common after epidemics have dissipated. The survivors are immune to the disease.

Types of disease statistics are as follows:

1. **Incidence rate**: the number of new cases of the disease in a population. This looks at the spread of the disease.

2. **Prevalence rate**: the total number of people infected at any point in time. Prevalence determines the duration of the disease.

3. **Morbidity**: the cases per 100,000 people per year.

4. **Mortality**: the deaths per 100,000 people per year.

The levels of disease in the population are described as follows:

1. **Endemic**—some are infected in a region all the time.

2. **Epidemic**—infection rate higher than normal. May have high morbidity and high mortality.

3. **Pandemic**—a worldwide epidemic.

4. **Sporadic disease**—a random and isolated case whose occurrence is unpredictable.

5. **Common source outbreak**—such as, everyone who eats a certain food getting food poisoning.

6. **Propagated epidemics**—direct contact between an infected person and a susceptible noninfected person.

There are several types of epidemiologic studies:

1. **Descriptive study**, which looks at the current disease epidemic and how it is spread.

2. **Analytical studies** compare activities of infected people to a control group from same population and find differences between them—possible source of the disease. Analytical studies are retrospective studies.

3. **Experimental studies** are often used to test treatments for diseases. In many experimental studies, the test group gets the test drug and a control group gets a placebo.

Topic Test 5: Epidemiology

True/False

1. A descriptive epidemiologic study is a study that compares treatments with placebos used by control groups.

2. Mortality of a disease refers to how may people per 100,000 of the population died from the disease.

Multiple Choice

3. A worldwide epidemic is called a(n)
   a. endemic.
   b. epidemic.
   c. pandemic.

4. Morbidity is
   a. the number of people who died from the disease.
   b. the number of cases of a disease at any point in time.
   c. the number of cases of the disease per 100,000 people per year.

Short Answer

5. What does a descriptive study of a disease look at?

Topic Test 5: Answers

1. **False.** A descriptive epidemiologic study looks at the current epidemic and how it spreads.

2. **True**

3. **c**

4. **c**

5. A descriptive study looks at who got the illness, how many people got sick, and the time and locations of the cases.

APPLICATIONS

Botulism toxin is reputed to be in Iraq's arsenal of biologic weapons. *Clostridium botulinum* toxin is also produced by several pharmaceutical companies in the United States. One company, Allergan, markets the botulism A toxin under the registered trademark of BOTOX. BOTOX is used to treat some neurologic disorders, such as dystonia, which is characterized by constant muscle contractions. Some people also use BOTOX to reduce their facial wrinkles.

DEMONSTRATION PROBLEM

Fifty-three children, students at X elementary school, developed diarrhea and high fever. All 53 children became ill during the first week in October. The children were from all grade levels and were scattered among all the classrooms. Enteropathogenic *E. coli* was cultured from stool specimens from all the ill children. Further analysis found that all 53 children always bought their lunches in the school cafeteria. However, 67 other children who also bought their lunches did not become ill.

A cafeteria worker became ill with the same symptoms during the same week. The cafeteria worker ate her lunch in the school cafeteria, too. The cafeteria worker ate a hamburger on Monday. The other options for entrees those days were hot dogs and macaroni and cheese. The children who ate in the cafeteria and did not become ill ate the macaroni and cheese and hot dogs. The common ingredient between the hamburgers and the spaghetti and meat sauce was hamburger purchased from a local packing house.

Solution

E. coli is likely to be found in other meat from the packing house.

Chapter Test

True/False

1. Horizontal transmission of a disease is from parent to offspring.

2. Virulence factors are what enable the pathogen to invade host cells or tissues.

3. Skin is one type of epithelial tissue.

4. Portals of entry for organisms include the natural body openings, such as the mouth, nose, tear ducts, and urethra.

5. Endotoxins are excreted by the invading bacteria.

6. The presence of pathogens in the body is abnormal.

7. Organisms that cannot attach to the host cells are pathogens.

Multiple Choice

8. Which term(s) describes the level of disease in the population?
   a. Endemic
   b. Epidemic
   c. Pandemic
   d. All of the above

9. Which category of pathogen causes disease only when it has the opportunity?
   a. Nonpathogen
   b. Opportunist pathogen
   c. Pathogen

10. Which of the following are types of disease statistics?
    a. Incidence rate
    b. Prevalence rate
    c. Morbidity
    d. Mortality
    e. All of the above

Short Answer

11. How do the enzymes categorized as virulence factors help the invading bacteria establish an infection?

12. What are the types of symbiotic relationships?

13. How do exotoxins differ from endotoxins?

14. What is a communicable disease?

15. What are antiphagocytic factors?

16. What type of toxin is *only* produced by gram-negative bacteria?

17. How can botulism toxin be destroyed?

Chapter Test Answers

1. **False**

2. **True**

3. **True**

4. **True**

5. **False**

6. **True**

7. **False**

8. **d**

9. **b**

10. **e**

11. Enzymes categorized as virulence factors enable the bacterium to establish an infection either by attacking the host cells and tissues to allow the organism to invade or by protecting the bacterium from the host defenses.

12. The types of symbiotic relationships are commensal, parasitic, and mutualistic.

13. Exotoxins differ from endotoxins in the following ways:
    a. They are excreted by the bacterium, whereas endotoxins are part of the cell wall;
    b. Exotoxins are proteins and endotoxins are lipopolysaccharides;
    c. Exotoxins are good antigens, and endotoxins are not;
    d. Exotoxins are easily destroyed by heat, and endotoxins are not;
    e. Exotoxins can be produced by gram-positive or gram-negative organisms, and endotoxins are only produced by gram-negative organisms;
    f. People are immunized against some exotoxins, but people cannot be immunized against endotoxins.

14. A communicable disease is one that can be spread before the infected person shows symptoms and after the infected person has recovered from the disease.

15. Antiphagocytic factors are one of the virulence factors that enable an organism to produce disease.

16. An endotoxin is only produced by gram-negative bacteria because it is part of the gram-negative cell.

17. Botulism toxin can be destroyed by heat.

Check Your Performance

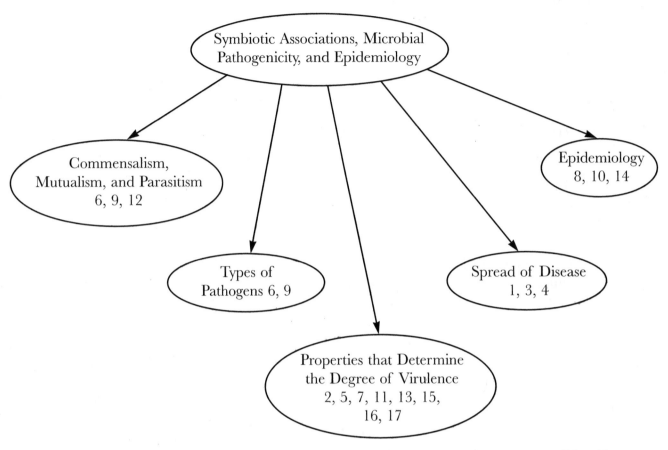

Note the number of questions in each grouping that you got wrong on the chapter test. Identify areas where you need further review and go back to relevant parts of this chapter.

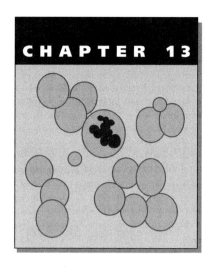

CHAPTER 13

Host Defense Mechanisms: Nonspecific and Constitutive

Large complex multicellular organisms, such as mammals or birds, daily face a barrage of microorganisms seeking to parasitize them, to exploit the nutrients accumulated, or to inhabit environmentally favorable sites on these organisms. Despite this barrage, "disease is the exception, not the rule." Multicellular organisms have developed a defense system consisting of three lines. This system has reserves in case the first line is breached. Skin and mucous membranes form the first line of defense. If this is breached, phagocytic cells and toxic compounds in the blood and tissue fluids are brought to bear on the invaders. If these weapons are insufficient, the last line of defense consisting of the antibody system with humeral antibody and cell-mediated destruction can be activated. Both the boundary layers and the bloodborne defenses are **constitutive**, which means constantly present. They are also **nonspecific**, responding to any foreign element interacting with the host. In contrast, the antibody system is highly specific, responding only to particular foreign substances (see Chapters 14 and 15).

ESSENTIAL BACKGROUND

- **Commensalism (Chapter 12, Topic 1)**
- **Parasitism (Chapter 12, Topic 1)**

TOPIC 1: PHYSICAL BARRIERS

KEY POINTS

✓ *How does the architecture of the body's boundary serve as a barrier?*

✓ *In what ways are these barriers breached?*

Although not directly involved in defense, nutrition must be noted. Without balanced nutrition, the substances needed to maintain physical, chemical, and biologic defensive barriers will be deficient, and those barriers will fail.

The skin and linings of the gastrointestinal tract, mouth, nose, ear, and vaginal tract are composed of very tightly interdigitated cells. Very few organisms can breach this tightly compact layer of cells. The outer layer of keratinized skin cells resists enzymatic attack. Shedding skin cells and washing the skin remove adhering organisms. Skin is subject also to dryness.

The respiratory tract has hairs that filter out particles larger than 5 micrometers in diameter. Mucous layer traps particles down to 1 micrometer diameter. Ciliated cells line the trachea and bronchi to push particles and mucus to the throat where they can be ejected through the mouth or swallowed.

Peristalsis by muscles surrounding the gastrointestinal tract pushes microorganisms on food particles or in the stool through the tract and out the anus, another method of flushing out microorganisms. Urination flushes organisms that may have entered the urethral opening from the urethra, and tears flush the eyes.

A very few parasites have enzymes that digest the intercellular matrix. These parasites can enter the resulting gaps between cells. Cuts or punctured skin provides an easier entry. Some microorganisms adhering to the skin or mucous membranes can irritate them. Inflammatory responses by the host are designed to kill the invader but often are just as lethal to host cells, especially wounded host cells. If the invading organisms are not killed, the dead host cells provide a breach the surviving microorganisms can cross.

Numerous commensal bacteria, the **normal microflora**, seldom produce problems on the exposed surfaces of the body. Commensals compete for space and nutrients with pathogenic microbes, preventing attachment of the pathogens. In addition, commensals stimulate host-clearing mechanisms and cross-protective immune factors.

Topic Test 1: Physical Barriers

True/False

1. Some defense mechanisms must always be present.

2. Balanced nutrition is necessary for defense.

Multiple Choice

3. From which of the following surfaces are microorganisms *not* regularly flushed or physically rubbed off?
   a. Eyes
   b. Vaginal tract
   c. Skin
   d. Gastrointestinal tract
   e. Urethral entrance

4. What is probably the most common and significant way the outer line of defense is breached?
   a. Digestion of keratin
   b. Digestion of intercellular matrix
   c. Puncture the skin
   d. Adherence to outer surface
   e. Production of odiferous acids

Short Answer

5. How is the architecture of the skin designed to provide a formidable physical barrier to mechanical penetration?

6. How do commensals enhance the biologic barriers hosts use against pathogens?

Topic Test 1: Answers

1. **True.** Humans and other complex multicellular organisms are coming constantly in contact with microbes that could parasitize them.

2. **True.** Maintenance of physical, chemical, and biologic barriers requires an extensive array of compounds and energy that the diet provides.

3. **b.** Although the vaginal tract is not usually flushed regularly, it has other barriers that are effective in protecting it.

4. **c.** Microbes do not need any specialized features to penetrate a wound in the skin. Essentially any microbe has the chance of entering through a break.

5. The skin has two architectural features that make it an excellent physical barrier. First, the tightly interlocking cells are sealed with an intercellular matrix and, second, the outer layers are impregnated with keratin.

6. Commensals enhance a host's biologic barriers by toxin production, occupying space, preventing pathogen attachment, consuming scarce nutrients, influencing host clearing actions, and inducing cross-reactive factors.

TOPIC 2: CHEMICAL BARRIERS

Key Points

✓ *Where are these chemical barriers found?*

✓ *How do the products of commensals sometimes aid hosts?*

Skin and mucous membranes are coated with a variety of toxic chemical substances. The high salt content of sweat increases osmolarity of water on body surfaces so few microorganisms can multiply. Sebaceous glands secrete fatty acids that are toxic to most microorganisms. Mucus contains lactoferrins, iron-binding proteins that chelate iron atoms tightly, starving many microorganisms for iron. Gummy mucus coats many microorganisms, creating anaerobic conditions or preventing attachment. Low pH of urine, vagina, and skin inactivate key enzymes of microorganisms. In the gastrointestinal tract, microorganisms must tolerate a strong acid bath in the stomach. Some succeed in passing through the stomach bound to food particles, only then to face digestive juices in the duodenum. The juices contain degradative enzymes and a lipid solubilizing detergent bile from the liver. Mucus of the respiratory system and the vaginal tract contain toxic products. Tears and saliva have lysozyme, an enzyme that degrades bacterial cell walls.

Blood and tissue fluids, the second line of defense, contain many chemicals adverse to bacteria. Fibronectin, a large glycoprotein, coats the surfaces of some bacteria, preventing their attachment to cells. Beta-lysin, leukins, plakins, and phagocytin are cationic polypeptides produced in platelets and leukocytes to disrupt cytoplasmic membranes of gram-positive bacteria. Lactoferrins that bind iron are also found in blood. The interferon family of glycoproteins alerts cells to viral infections and triggers those cells to inhibit viral nucleic acid synthesis. Nonspecific antibody-like proteins tag foreign organisms, enhancing their ingestion by phagocytes.

Finally, commensals living on the body surfaces, but doing no harm, may produce substances that add to the chemical barriers provided by the host. Commensals found in the large intestine produce metabolic products, such as fatty acids and hydrogen sulfide. Members of the bacterial family Enterobacteriaceae produce bacteriocins that lyse other bacteria. These organisms create anaerobic conditions that are unfavorable to many pathogenic bacteria.

Topic Test 2: Chemical Barriers

True/False

1. Various cells of the host produce all the chemicals used to combat foreign organisms.

2. Some chemicals rather than killing create nutrient starvation, such as iron-binding lactoferrins.

Multiple Choice

3. Which of the following is *not* part of the chemical barrier found in the gastrointestinal tract?
   a. Fatty acids
   b. Strong acidity
   c. Degradative enzymes
   d. Detergent liver bile
   e. High salt level

4. Tears and saliva contain which of the following antibacterial agents?
   a. Fatty acids
   b. Lysozyme
   c. Beta-lysins
   d. Interferon
   e. Fibronectin

Short Answer

5. How do commensal microorganisms help multicellular hosts combat pathogenic bacteria and fungi?

Topic Test 2: Answers

1. **False.** Some elements in chemical barriers arise from the metabolism of host products by commensal microorganisms.

2. **True.** Many pathogenic bacteria have a requirement for iron. Lactoferrins bind iron tightly, creating an iron deficiency in the vicinity of the invading bacteria.

3. **e.** The high salt levels occur on the skin as a result of the evaporation of sweat.

4. **b.** The other substances listed occur on the skin and mucous membranes or in blood and tissue fluids.

5. Commensal organisms metabolize host compounds to produce acids, fatty acids, or hydrogen sulfide that may be toxic to many pathogens.

TOPIC 3: BIOLOGIC BARRIERS

KEY POINTS

✓ *In what ways do the cells of multicellular hosts create barriers against pathogenic microorganisms?*

✓ *In addition to producing toxins, how else do commensal microorganisms create defensive barriers for their host?*

Biologic barriers represent the heart of the second line of defense, sometimes called the **reticuloendothelial system**. Several types of phagocytic cells are scattered throughout the body to injest and degrade foreign substances. Some are motile, some are fixed, and some may move from one state to the other. **Fixed macrophages** are bound to a network of protein fibers or embedded in the linings of passages through the **lungs**, **liver**, **spleen**, and **lymph nodes**. As fluid flows through these filter organs, the fixed macrophages extend pseudopods to engulf any passing foreign material. Circulating in blood are phagocytic **monocytes** and two types of phagocytic **granulocytes**. Granules in the granulocytes are lysosomes filled with lytic enzymes that fuse with vacuoles containing foreign material. **Neutrophils** are directed against smaller particles. Lysosomal enzymes of **eosinophils** damage the cell membranes of large multicellular parasites. Granulocytes arrive at the scene of an invasion first. Monocytes are used as reserves in case granulocytic activity is insufficient.

Phagocytosis begins by contact with the foreign agent, for example a bacterium, then engulfment of that agent. The cytoplasmic **phagosome** containing the foreign agent fuses with a lysosome to form a **phagolysosome**. Released lysosomal enzymes digest the foreign agent in the phagolysosome. The detritus-filled phagolysosome moves to the cytoplasmic membrane, fuses with it, and dumps the detritus by **exocytosis.** These processes produce **oxygen-independent** killing. Coupled to these actions is a second **oxygen-dependent** series of enzymatic reactions. Because oxygen is required, the phagocytic cells experience a **respiratory burst** producing very strongly oxidative substances. Neutrophils synthesize additional oxidative compounds.

Wounded cells and clotting factors initiate an **inflammatory response**. Damaged cells and breaks in the vascular system indicate the body's outer defenses have been breached. Defenses are activated in case an invasion has occurred. The defense system removes damaged and dead cells so healing can progress. **Inflammation** triggers localized dilation of blood vessels near the wound. Fluid accumulates in the tissues, causing swelling. Blood flow slows, allowing for accumulation of phagocytic cells and oxygen for the respiratory burst. Cells in the wound area release certain **cytokines** to trigger chemotaxis of leukocytes to the area and inhibit their migration. Other substances released increase vascular permeability, permitting arriving neutrophils and monocytes to migrate into the tissues to the injury. Emitted pyrogens increase cellular metabolism and temperature (fever). Because most pathogenic organisms live near their thermal death point, a few extra degrees in host tissue may eliminate invaders.

Topic Test 3: Biologic Barriers

True/False

1. For prophylactic reasons, whenever cells are wounded or the vascular system cut, an imflammatory reaction will be initiated.

2. Once engulfed in a vacuole of a phagocytic cell, a foreign organism faces a lethal barrage of highly oxidative substances.

Multiple Choice

3. Which of the following is *not* a site in humans where invading microorganisms are filtered from bodily fluids passing through the site?
   a. Liver
   b. Heart
   c. Lung
   d. Lymph node
   e. Spleen

4. Which of the following cells, although involved in defense, is *not* phagocytic?
   a. Eosinophils
   b. Neutrophils
   c. Lymphocytes
   d. Fixed macrophages
   e. Monocytes

Short Answer

5. How does inflammation prepare for a possible entrance of a foreign organism?

Topic Test 3: Answers

1. **True.** These conditions usually indicate the first line of defense has been breached and it is likely foreign organisms will enter.

2. **True.** Upon phagocytosis, phagocytic cells undertake a respiratory burst in which highly oxidative compounds are synthesized.

3. **b.** Putting a meshlike filtering network in a pump like the heart would probably be a fatal design flaw.

4. **c.** Although lymphocytes through their production of antibody are very involved in defense, they are not phagocytic.

5. Inflammation produces concentration of leukocytes, increasing vascular permeability for leukocytes, increasing regional temperature above the thermal death point of many potential invading microorganisms.

Most commensal organisms on the body's surfaces are opportunistic pathogens. If they penetrate the body's outer defenses, they can institute very serious infections. These organisms lack mechanisms to actively penetrate the body's outer boundaries. A case in point is the widespread organism *Staphylococcus aureus*. In surgical wards of hospitals, this organism can penetrate the wounds deliberately created in the surgical procedure. Surgical procedures also tend to weaken a patient's immune system (the third line of defense discussed in Chapters 14 and 15). This necessitates that sterile techniques are used around the patient to minimize contamination. Resistance to antibiotic therapy is a significant problem with *S. aureus* infections. *S. aureus* can be a problem for unclean individuals because large populations can develop in clogged sweat ducts and penetrate the cells lining the duct. Mechanical scouring most of us use when washing our skin removes most of the organisms, keeping the population size of *S. aureus* in check.

Chapter Test

True/False

1. Skin and mucous membranes present a formidable line of defense because they incorporate physical, chemical, and biologic barriers.

2. For optimal health, body surfaces should be disinfected frequently to maintain sterility.

Multiple Choice

3. Which of the following are mechanisms used to prevent colonization from the gastrointestinal tract?
   a. Peristalsis
   b. Strong acid
   c. Microbial competition
   d. Degradative enzymes
   e. Detergent action of liver bile
   f. All of the above
   g. b, d, and e.

4. Prevention of microbial colonization of lungs is accomplished by which of the following?
   a. Phagocytic cells
   b. Filtration by hairs
   c. Pushing of ciliated cells
   d. Gummy mucus
   e. Coughing reflex
   f. All of the above
   g. b, c, and d.

5. What activity of the commensal *Lactobacillus acidophilus* is encouraged in the vaginal tract?
   a. Production of a capsule
   b. Creation of anaerobic environment

c. Conversion of glycogen to lactic acid

d. Production of toxic odors

e. Stimulation of inflammatory response

Short Answer

6. What are constitutive defenses and what is the significance of them?

7. What is probably the most significant feature commensals offer in the defense of their host?

Essay

8. Unfortunately, the chicken you have eaten was improperly cooked, and *Salmonella* has been introduced into your mouth. What defenses must these bacteria overcome before they can colonize your intestine?

Chapter Test Answers

1. **True**

2. **False**

3. **f**

4. **f**

5. **c**

6. Constitutive defenses are constantly present and are nonspecific, being able to function against any invader. Because the external surfaces of a host are exposed continually to microorganisms, some defenses are needed continuously.

7. Commensals occupy space, making it difficult for invading microorganisms to attach to a host's surfaces. In addition, large populations of commensals consume available nutrients, leaving newly arriving organisms to starve.

8. First, the *Salmonella* must not be sensitive to the carbohydrate degrading enzymes in saliva nor can they be trapped in any mucus from the throat. Upon entering the stomach, the *Salmonella* must hide on or around food particles that neutralize the acid there. As the *Salmonella* enter the duodenum of the small intestine, they must tolerate a sudden 10,000-fold decrease in acidity, have a protective coat to resist pancreatic digestive enzymes, and be impervious to liver bile. As they move further through the small intestine, they must cling to the intestinal wall or get flushed out by peristaltic activity. There they meet macrophages ready to engulf them and must survive digestion by lysosomal enzymes. Also, they must successfully compete against many commensals already present. Surprisingly, some manage to do all these things. One then develops food poisoning from the *Salmonella*.

Check Your Performance

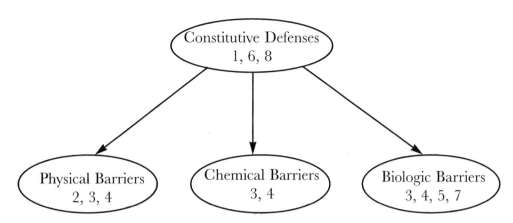

Note the number of questions in each grouping that you got wrong on the chapter test. Identify areas where you need further review and go back to the relevant parts of the chapter.

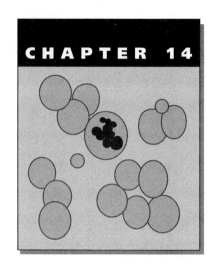

CHAPTER 14

Host Defense: Characterization of Antigens and Antibodies

In this and the next chapter, the inducible (acquired) host defense system or the **antigen-antibody system**, the third line of defense, will be explored. **Lymphocytes** produce **antibodies**, some of which remain associated with the cell that synthesized them and others that are released into the blood. An **antibody** is a protein having at least two sites that bind specifically with a structure on the surface of an antigen. An **antigen** is a foreign substance that activates certain specific clones of lymphocytes to produce antibodies that in turn react specifically with that antigen. There is a high degree of specificity between an **antigen** and an **antibody**. This **immune system** is found only in mammals and birds.

ESSENTIAL BACKGROUND

- **Protein structure (Chapter 2, Topic 3)**
- **Protein synthesis (Chapter 10, Topic 1)**

TOPIC 1: PROPERTIES OF ANTIGENS AND ANTIBODIES

KEY POINTS

✓ *What kinds of foreign materials can serve as antigens?*

✓ *How does the structure of an antibody result in the highly specific interaction between antigen and antibody?*

Antigens function in two ways: first as an **immunogen** they stimulate synthesis of antibody by host lymphocytes and second as an **antigen** they bind chemically to the resulting antibody. Certain characteristics of a foreign substance determine whether that substance will be **immunogenic**. Only small parts of the **immunogen** actually are involved in the immune response of the host. The degree of foreignness from host components makes it less likely that a common potentially immunogenic site is shared. Shared sites will be treated as **self** by the host's immune system, blocking synthesis of antibody to these sites. Molecular size, particularly of molecular weights around 100,000 daltons, are best. Molecules of less than 20,000 daltons are poor immunogens. Chemical composition and heterogeneity of the immunogen produce small distinctive projections on the surface of the antigen. An antigen may have several different types of immunologically active sites, **epitopes**, depending on its chemical composition. Each different **epitope** will stimulate a different clone of lymphocytes. The more heterogeneous the **epitopes** on an antigen are the greater the immunogenic response to that antigen because many different antibodies are produced and directed against that antigen. An **epitope** independent of its antigen is a **hapten** that can react with an antibody but induces no immunologic response because of its small size. Haptens are also small molecular weight molecules that can chemically react with a large molecular weight molecule, such as a protein, and become an epitope. Generally speaking, antigens are carbohydrates, proteins, or nucleic acids. Lipids rarely are immunogenic, probably because they are smaller than 20,000 daltons and have hydrophobic interactions. Immunogenic stimulation relies mainly on hydrophilic interactions. Exposure to low concentrations or high concentrations of antigen may induce **low zone** or **high zone tolerance** in the host where the host fails to produce an immunologic response even though an antigen is present.

Antibodies are proteins produced by a specific clone of B or T lymphocytes (see Topic 4) when activated by the presence of a particular **antigen**. Each **humoral** antibody molecule consists of four protein chains held together by disulfide bonds in the shape of the letter Y (**Figure 14.1**). The two **light chains**, κ and λ, are short and are 25,000 daltons. An antibody molecule has identical light chains. The two **heavy chains** are 50,000 daltons and are longer. Heavy chains of which there are five types (α, γ, μ, δ, and ε) are identical and determine the response elicited when an antigen binds to the antibody. Light chains have a **variable** and a **constant** domain (**Figure 14.2**). The **variable** domain forms part of the **antigen-binding site** and varies in its amino acid sequence from one antibody molecule to another. The **constant** domain has the same amino acid sequence from one antibody molecule to another except the constant region of κ differs from that of λ. Heavy chains have one **variable** domain and three or four **constant** domains. The **variable** domain forms the other part of the **antigen-binding site**. The **constant** domains determine the type of antibody, IgA, IgG, IgM, IgD, or IgE (Ig for immunoglobulin). Within the variable region certain areas that directly contact the antigen in the binding site are **hypervariable**; these areas in the genes coding for antibody proteins experience high rates of mutations (see Topic 5). The variable domains of antibody molecules give them their high degree of specificity toward antigen molecules following the image of a **lock** and **key**. At least

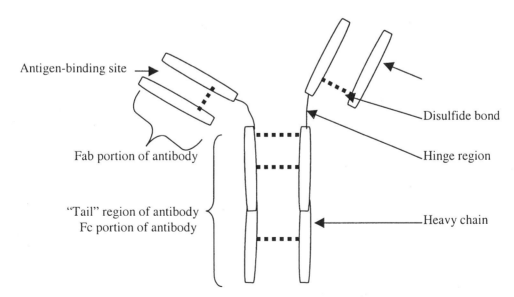

Figure 14.1. Diagram of a "typical" antibody molecule.

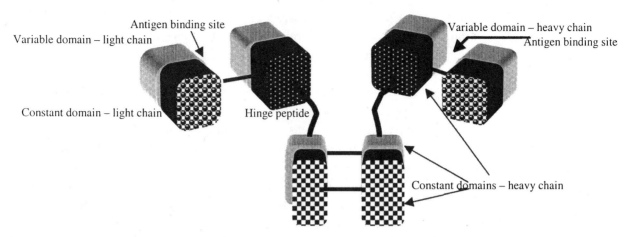

Figure 14.2. Variable and constant domains of a "typical" antibody molecule.

two binding sites occur on every antibody molecule. IgM is a polymer of five monomers held together by disulfide bonds and a J (joining) protein. IgA usually occurs as a dimer held together with J protein. Secretory IgA, a dimer complexed to secretory protein, is released onto body surfaces in tears, sweat, and milk.

When an antibody binds to an antigen, the constant domains of the heavy chains will activate one of four actions:

1. Neutralization of an antigen, such as blocking a toxin or attachment of a virus to a host cell membrane;

2. Opsinization (enhancement) of phagocytosis of an antigen by phagocytic cells;

3. Fixation of complement proteins leading to rupturing of cell membranes of foreign and nearby host cells;

4. Activation of release of cytotoxins by T_C lymphocytes (**T cytotoxic cells**);

When T lymphocytes synthesize antibody, those antibody molecules remain associated with the cell surface. The constant heavy chain domains are embedded in the T lymphocyte membrane, whereas the antigen-binding sites face the cellular environment.

Topic Test 1: Properties of Antigens and Antibodies

True/False

1. The lock-and-key model describes very well the binding between an antigen and an antibody.

2. Lipids make unusually good antigens because of their hydrophobic interactions with antibodies.

Multiple Choice

3. Which of the following reduces the immunogenicity of an antigen?
   a. Susceptibility to antigen processing and presentation by phagocytic cells
   b. Small molecular size; less than 20,000 daltons

c. Heterogeneity of chemical composition of the antigen

d. Foreignness compared with the host

e. All of the above

4. Which of the following does *not* happen when an antigen and an antibody react together?

a. Activation of complement

b. Release of cytotoxins from T_C lymphocytes

c. Opsonization of phagocytosis

d. Neutralization of antigen

e. Thermal denaturation of antigen

5. Immunologically active site on an antigen is what?

a. Heavy chain

b. Epitope

c. Light chain

d. Disulfide bridge

e. Constant domain

Short Answer

6. What features of the antibody molecule make it so specific for a certain antigen?

Topic Test 1: Answers

1. **True.** The very close fit of the epitope into grooves in the antigen-binding site resembles the shape of a key that fits into the tumblers on a particular lock.

2. **False.** Just the opposite is true. Antibodies use hydrophilic forces to bind antigens.

3. **b.** Immunogenic antigens must be greater than 20,000 daltons; in fact, 100,000 daltons is best.

4. **e.** Although fever does play a role in inhibiting parasites, antigen-antibody reactions are not involved.

5. **b.** Epitopes are projections on the surface of antigens that both trigger antibody production and antibody binding.

6. The variable domains of both heavy and light chains permit millions of different combinations. Hence, the antigen-binding site can have an enormous number of shapes to fit around different epitopes occurring on antigens.

TOPIC 2: PROPERTIES OF THE IMMUNE SYSTEM

KEY POINTS

✓ *In what ways does the immune system protect hosts from harm?*

✓ *How is self- not self-recognition achieved?*

Acquired immunity specifically recognizes and selectively eliminates foreign antigens. When the immune system is activated, it must react strongly, be destructive to foreign substances, and become quiescent when the foreign substances are eliminated. To prevent its destructive powers from being unleashed against the host, potential host antigens must be recognized and tolerated.

The immune system must have **antigenic specificity**. Specificity resides in the activation of specific clones of lymphocytes that produce the antibodies with binding sites that interact with very specific **antigenic sites** (**epitopes**).

The immune system must have **diversity**. Millions of foreign antigens exist in nature, and the immune system must recognize and eliminate each one specifically. A lymphocyte produces only one type of antibody molecule. Therefore, the immune system must contain millions of different lymphocytes. How can the host make all these antibodies without committing all its genes to coding for antibodies? Surprisingly, only about 300 to 400 genes are necessary (see Topic 5).

The immune system provides **immunologic memory** (anamnestic response). Second encounters lead to about a fivefold increase in detectable response compared with the initial encounter with a specific foreign substance. Memory cells are produced as well as antibody-synthesizing cells when a foreign antigen is encountered.

The immune system has a mechanism for preventing self-destruction. **Self-/not self-recognition** is established shortly before birth in mammals or hatching in birds. During a narrow window of time any antigen, whether native to the host or introduced at that time, will be considered **self**. Apparently, any clones of lymphocytes recognizing these **self** antigens will be destroyed. Hence, the immune system has no means of mounting a response to self antigens. Any antigen experienced after this window of time during development will be recognized as **not self**, and a reaction made against it. **Self-/not self-recognition** provides for the elimination of any new antigen whether formed by a host cell or introduced from the environment. The immune system can remove mutated host cells, some of which may be capable of uncontrolled cell multiplication (a cancer). Because cancer cells usually form unique tumor antigens, **cancer cell surveillance** can be based on self-/not self-recognition.

Topic Test 2: Properties of the Immune System

True/False

1. Self-/not self-recognition provides mechanisms so the immune system will not destroy the individual.

2. Acquired immunity serves to focus all defense processes against a limited number of foreign substances.

3. A particular clone of lymphocytes makes different antibodies depending on the foreign substance in the host.

Multiple Choice

4. Immunologic memory is demonstrated when
   a. a large amount of antigen elicits no response.
   b. a protein sharing an epitope with a digestive enzyme is tolerated.
   c. second exposure to a bacterium triggers a fivefold faster antibody production.

d. three different antibodies are directed against a single antigen.

e. None of the above

5. The inducible immune system must

a. be destructive of foreign substances.

b. react intermittently.

c. produce antibiotics.

d. remain constitutive.

e. produce antibodies against self antigens.

f. All of the above

Short Answer

6. Why is cancer cell surveillance linked to self-/not self-recognition?

Topic Test 2: Answers

1. **True.** This is probably done by destroying lymphocytes that would be activated by host antigens.

2. **True.** Because one or, at the most, a few foreign substances invade a host beyond the capacities of the constitutive defenses, activation of acquired immunity with its specificity allows channeling all the host's defense mechanisms against those few invaders.

3. **False.** A clone of lymphocytes produces only one type of antibody.

4. **c.** The acquired immune system seems to "remember" the foreign antigens to which it has had previous exposure and responds more rapidly.

5. **a.** Antibiotics are toxins usually produced by bacteria or fungi. Destroying self antigens would lead to self-destruction. Fast response and damping down must occur also.

6. Because cancer cells deposit substances new to an individual on their surfaces, these substances will be recognized as not self, and the acquired immune system will target their destruction before serious physiologic dysfunction occurs.

TOPIC 3: TYPES OF IMMUNITY (ACTIVE AND PASSIVE)

KEY POINTS

✓ *What distinguishes active from passive immunity?*

✓ *Why is inducing active immunity an advantage to humans?*

On first exposure to an antigen, only a few of an individual's B lymphocytes respond to that particular antigen (see Topic 1). The responding B lymphocyte is highly specific to the triggering antigen. Activated B lymphocytes begin dividing, forming two types of progeny cells. Most progeny cells will synthesize antibody directed specifically against the triggering antigen. A smaller number of these cells will remain inactive and form a population of **memory** cells that can be activated on a subsequent exposure to this same antigen. These cells held in reserve form

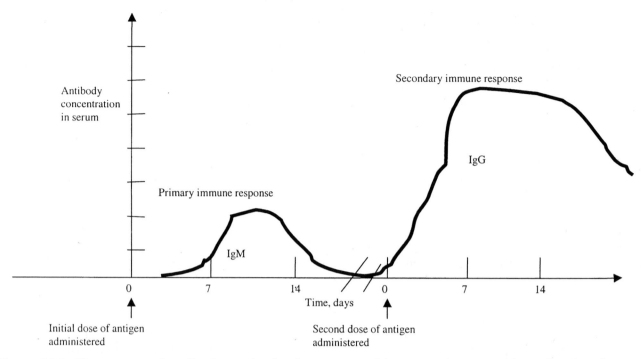

Figure 14.3. Time course of antibody production by a mammal (anamnestic response). Notice that faster and greater response after the second exposure to the antigen, indicating some method of immunologic memory.

the phenomenon known as **immunologic memory**. On first exposure, 10 to 14 days are required before measurable quantities of antibody to the antigen are detectable (**Figure 14.3**). Detectable antibodies levels will be present in 3 days on subsequent exposures to this antigen because of the buildup of memory cells. **Primary responses** are short-lived, **secondary responses** last for a protracted time, sometimes years.

The immune system can be used in therapy or for prophylaxis against agents or organisms that might be extremely harmful. Surprisingly, antibodies produced by one mammal are effective not only in that animal but also in mammals of the same or other species. This can be exploited if an individual is exposed to an acutely toxic substance(s). For example, a rattlesnake bites an individual. One does not have 14 days to produce sufficient anti-rattlesnake antibodies before the rattlesnake venom has caused severe injury or fatality. Injecting the victim with horse serum containing anti-rattlesnake antibodies can neutralize the venom, reducing damage from the venom. This short-lived procedure is called **passive immunity** because the victim did not synthesize the antibodies but received them from another organism. Repeated use may stimulate the recipient to produce anti-antibodies, leading to a reaction potentially as disastrous as the venom in our example.

Active immunity or **vaccination** achieves long-term protection for prophylaxis. Active immunity occurs when an individual is deliberately exposed repeatedly to an antigen. This individual's immune system will establish immunologic memory that can be activated rapidly if the individual is exposed to that antigen. For example, a widespread soil bacterium *Clostridium tetani* can enter wounds, multiply sparingly, and produce a potent neurotoxin. **Vaccinations** can protect individuals from several serious diseases, such as polio, diphtheria, whooping cough, measles (rubiola), mumps, German measles (rubella), chicken pox, and hepatitis A and B.

Topic Test 3: Types of Immunity

True/False

1. Passive immunity occurs when a population of memory cells is formed but no antibody is formed.

2. There is a high degree of specificity between an antigen and the antibody made in response to the presence of that antigen.

Multiple Choice

3. Immunologic memory occurs after vaccination because
   a. antigens in the vaccine are chemically modified.
   b. the host produces respiratory toxins against the antigen.
   c. clones of lymphocytes capable of producing antibody against the antigen are increased.
   d. long-lived antibody molecules are produced by antigen-sensitive lymphocytes.
   e. None of the above

4. Active immunity can be used prophylactically to prevent which of the following diseases?
   a. Tetanus
   b. Polio
   c. Mumps
   d. Measles
   e. Hepatitis B
   f. All of the above

Short Answer

5. Why is passive immunity used only in situations of acute conditions?

Topic Test 3: Answers

1. **False.** Passive immunity occurs when a patient is given antibodies made in another host against antigens of an agent that is an acute threat.

2. **True.** The immune system is designed to respond against a specific foreign substance, the antigen. In a way the antigen specifically triggers its own destruction.

3. **c.** By increasing the population of a clone of lymphocytes, the patient is able to produce more antibody against an antigen more quickly.

4. **f.** Because of vaccination these typically childhood diseases are rare in the industrialized world. Major vaccination campaigns worldwide by the World Health Organization may extend this fact throughout the world shortly.

5. Because the patient is receiving antibody produced in another organism usually only once, the donated antibodies will be used or removed after a short period in circulation. The patient is not producing replacement antibody molecules. Passive immunity is used when a

patient has been exposed to a foreign material that would elicit acute physiologic distress before the patient had time to recruit his or her immune system.

TOPIC 4: BIOLOGY OF THE LYMPHOID SYSTEM

KEY POINTS

✓ *What distinguishes T and B lymphocytes?*

✓ *What are the differences between primary and secondary lymphoid structures?*

Hematopoiesis, production of new blood cells, occurs in bone marrow. There, stem cells differentiate into several blood cell lines. One line gives rise to lymphocytes. These lymphocytes are said to be **virgin**, meaning they are not **immunocompetent**. They must undergo selection and processing into either a T lymphocyte or a B lymphocyte by migrating via blood to the thymus gland or bursal regions in bone marrow. Selection is necessary to ensure that those which would produce antibodies against self antigens are eliminated. Those lymphocytes programmed to produce anti-self antibodies suffer apoptosis (suicidal cell death) in both areas where T or B lymphocytes mature.

T lymphocytes mature in the **thymus gland** located just above the heart and divided into an outer cortical layer and a central medullary region. The cortex is populated by a dense concentration of immature T lymphocytes (thymocytes). Amidst the thymocytes are scattered **nurse cells** that extend processes around many thymocytes. Apparently, nurse cells trigger the deposition of new surface components and the removal of existing ones from the thymocytes. Maturing T lymphocytes migrate to the medulla of the thymus gland to finish maturation into one of three types: T_H, T_C, or T_S. Mature T lymphocytes leave the thymus gland via blood.

In the **bursal** areas of bone marrow, a second line of lymphocytes, **B lymphocytes**, undergoes "maturation." Birds have an organ, **bursa of Fabricious**, near the end of the large intestine where their B lymphocytes mature. Less is known about B lymphocyte maturation, but there are changes in the components on the surface of B lymphocytes during maturation. After maturation, B lymphocytes leave bursal areas of bone marrow via blood.

Formation and maturation of lymphocytes occurs in **primary** lymphoid tissues (bone marrow, thymus, and bursal areas). When released in blood, mature lymphocytes migrate to lymph nodes and spleen, **secondary** lymphoid tissues. Small numbers of lymphocytes remain in circulation. Lymph nodes are organized in layers with B lymphocytes localized in follicles and germinal centers of the cortical layer and T lymphocytes concentrated in the paracortex (**Figure 14.4**). Medulla is populated mainly with **plasma cells**, terminally differentiated B lymphocytes actively producing and releasing humeral antibody. Phagocytic **macrophages** and **dendritic cells** are scattered through the cortex and paracortex. Dendritic cells with their long thin projections phagocytize soluble antigens, whereas macrophages phagocytize particulate antigens. Spleen can be considered a giant lymph node. When clones of lymphocytes are activated, they divide and form prominent germinal centers of progeny cells (see Chapter 15, Topic 1). As cell numbers increase, the lymph node swells to accommodate the growing volume of cells.

Mucosal-associated lymphoid tissue (**MALT**) in clusters associated with mucosal surfaces and **cutaneous-associated lymphoid tissue** found in the epidermis of the skin remove foreign organisms before they can penetrate the body very far. Phagocytic cells populate these

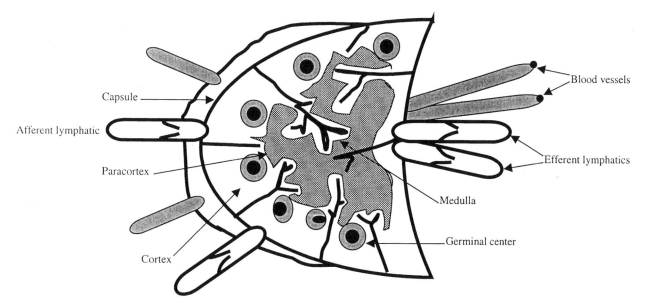

Figure 14.4. Diagram of a "typical" lymph node showing the significant structural features.

tissues, trap the foreign material, and migrate to the nearest lymph node where lymphocytes are activated.

Topic Test 4: Biology of the Lymphoid System

True/False

1. Virgin lymphocytes differentiate from stem cells located in the lymph nodes.

2. Lymphocytes must be processed and mature in bursal areas of bone marrow or the thymus gland before they become immunologically competent.

Multiple Choice

3. When B lymphocytes finish their maturation, they migrate to
   a. lung.
   b. heart.
   c. liver.
   d. lymph nodes.
   e. gall bladder.

4. Associated lymphoid tissue found in skin and mucous membranes performs what function?
   a. Provide maturing lymphocytes with surface components
   b. Phagocytosis of foreign antigens
   c. Selection of clones of lymphocytes to purge
   d. Accept antibody from B lymphocytes
   e. Convert to T lymphocytes in skin and mucous membranes

Short Answer

5. Why are clones of lymphocytes programmed to produce antibody against self antigens not observed in lymph nodes?

Topic Test 4: Answers

1. **False.** Virgin lymphocyte differentiation occurs in bone marrow.

2. **True.** Cell surface components are gained or lost during processing and maturation before lymphocytes can be activated to produce antibody.

3. **d.** Lymph node is where competent lymphocytes are warehoused until processed foreign antigens activate clones of them.

4. **b.** Phagocytized foreign material is brought to the nearest lymph node to activate the proper clones of lymphocytes to synthesize antibody.

5. During the maturational events in the thymus gland or bursal tissue, those clones of lymphocytes destined to produce anti-self antibody commit suicide (apoptosis) by an unknown mechanism.

TOPIC 5: GENETICS OF ANTIBODY FORMATION AND DIVERSITY

KEY POINTS

✓ *How is diversity of millions of unique antibody molecules achieved without relegating the entire genome to antibody production?*

Organisms that produce antibody are able to generate millions of different unique antibodies. Because each antibody is a protein, the Watson-Crick paradigm for protein synthesis would require a gene for each unique antibody or on the order of a million genes (Chapter 10, Topic 1). In an organism, such as a human, who has enough DNA for 100,000 to one million genes, every available gene would be used for antibody synthesis. Obviously, this is not happening. Eukaryotic genes are discontinuous, having **exons**, regions of information, split by **introns**, regions apparently genetically silent. This arrangement of informational genetic sequences opens up two additions for genetic variation: **mitotic recombination** of germline DNA during differentiation of the lymphocyte from the bone marrow stem cell and **RNA splicing** during transcription to form mRNA (**Figures 14.5** and **14.6**). **Mitotic recombination** in eukaryotic cells opens the possibility of much variation without many genes if **exons** can be recombined in many ways.

The κ light chain gene of humans is divided into four segments that must be brought together to make a useful mRNA. Through a combination of mitotic recombination and RNA splicing, an L component, a leader for targeting the antibody; a V component, containing variable sequences; a J component, also part of the variable domain; and a C component, the constant domain, must be put together. For human DNA it is estimated there are one L sequence, $100 V_\kappa$ sequences, $5 J_\kappa$ sequences, and one C_κ sequence. If these sequences can be combined in any order, there are 500 possible final mRNA sequences for coding κ light chains, requiring 107 genes. A similar situation exists for the human λ light chain except there are six J regions,

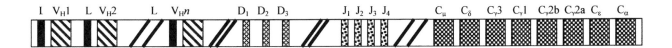

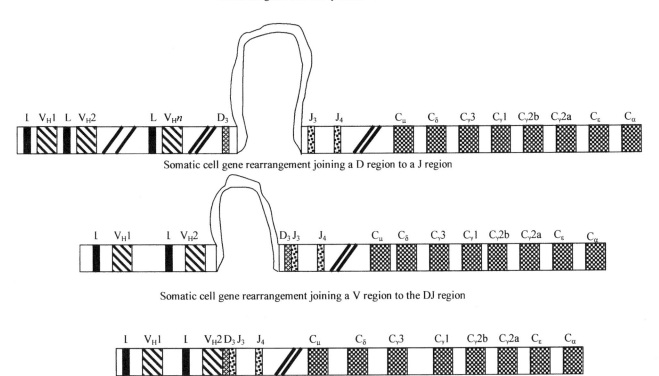

Exons for germ-line heavy chain DNA

Somatic cell gene rearrangement joining a D region to a J region

Somatic cell gene rearrangement joining a V region to the DJ region

Rearranged heavy chain DNA

Figure 14.5. Gene rearrangements among exons of an immunoglobulin heavy chain gene in an immature B lymphocyte.

meaning there will be 600 combinations for λ chains. Associated with every light chain in an antibody is a heavy chain. To synthesize a heavy chain, an L component must be spliced to V, D, J, and C components. For a human heavy chain there are estimated to be 1 L sequence, $100 \, V_H$ sequences, $30 \, D_H$ sequences, $6 \, J_H$ sequences, and 1 C sequence (Figure 14.5). If any combination can form, there will be 18,000 possible heavy chains, using only 138 genes. But the big gain in diversity comes because apparently any heavy chain can be combined with any light chain. **Combinatorial association** would result in 18,000 heavy chains combined to $500 \, \kappa$ light chains for 9,000,000 types of antibody molecules plus another 10,800,000 antibody types from heavy chain combined to λ light chains. Thus, one finds a **minimum** of nearly 20,000,000 possible unique antibody molecules with the use of only 353 gene sequences.

Topic Test 5: Genetics of Antibody Formation and Diversity

True/False

1. Every unique antibody molecule requires a unique stretch of DNA to hold its information.

2. Mitotic recombination and RNA splicing can increase diversity rapidly.

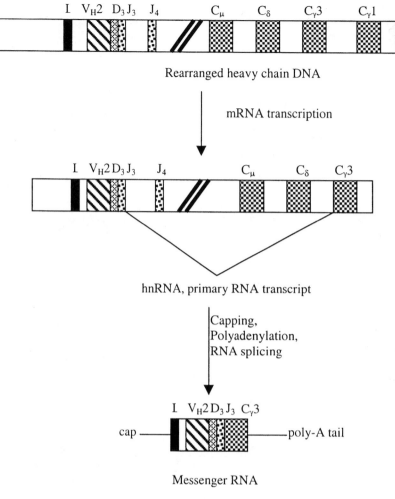

Figure 14.6. Formation of messenger RNA for an IgG molecule in a B lymphocyte.

Multiple Choice

3. The information for eukaryotic proteins is found in what DNA sequences?
   a. Introns
   b. Cistrons
   c. Replicons
   d. Exons
   e. Hesitons

4. Approximately how many unique antibody molecules must one be able to make to protect against foreign substances?
   a. 10^4
   b. 10
   c. 10^6
   d. 5×10^7
   e. 1

Short Answer

5. What does combinatorial association mean?

Topic Test 5: Answers

1. **False.** Through the use of mitotic recombination and RNA splicing, human lymphocytes can generate 20 million unique antibody molecules, using only 353 DNA sequences.

2. **True**

3. **d.** Introns are quiet DNA sequences separating exons; cistrons are the DNA sequences coding for a single protein; replicons are units of DNA replication; hesiton is a made-up word.

4. **c.** At least one million unique antibodies are needed to ensure protection against foreign substances.

5. Combinatorial association is the possible interaction of any heavy chain a lymphocyte might synthesize with any light chain the cell might synthesize. This process adds the most diversity of all to producing unique antibodies.

APPLICATION

When human immunodeficiency virus-1 (HIV-1) infections emerged as acquired immunodeficiency disease (AIDS) in the late 1970s, hope was held that fairly quickly a vaccine could be developed that would prevent infection. Because many antiviral vaccines lead to neutralizing antibodies against protein coats, blocking viral attachment to host cells, attempts were made to develop vaccines using the coat proteins from an isolate of HIV-1. In vitro testing with this vaccine proved unsuccessful. Further study of HIV-1 protein coats found them to be very diverse, even from a single starting clone of virus. Part of this variability can be attributed to high mutation rates. Perhaps a much more significant source of variation in HIV-1 is the "diploid" nature of the RNA (two RNA strands in each virus) unique to the retroviruses of which HIV-1 is. If these two RNA strands differ, they can undergo a recombination, leading to continual generation of varying protein coats. This makes it very difficult to prepare a vaccine that will be effective.

Chapter Test

1. The variable domains of heavy and light chains of antibody molecules permit the highly specific binding of an antigen.

2. B lymphocytes are generated in bone marrow and migrate to bursal regions in bone marrow where they mature.

3. T lymphocytes migrate via the blood to the liver after processing and maturation in the thymus gland.

4. Because a lymphocyte can produce antibodies against several different epitopes, one can understand easily how millions of antibodies, each specifically directed against a different epitope, can be produced.

5. The greatest contribution to antibody diversity arises during the combinatorial association of heavy chains with light chains that occurs during antibody production in a lymphocyte.

Multiple Choice

6. When the immune system is activated, which of the following responses does the immune system undertake?
   a. Produces destructive substances against the foreign material
   b. Reacts strongly
   c. Dampens down when the foreign material is removed
   d. All of the above
   e. None of the above

7. Which of the following is *not* an action generated by antigen-antibody binding?
   a. Block attachment of viruses to host cells
   b. Activate T_C cells to release cytotoxins
   c. Trigger complement proteins to bind to antigen
   d. Neutralize foreign toxin activity
   e. Enhance phagocytosis of antigen by macrophages
   f. None of the above

8. Diversity is introduced in which of the following processes during antibody formation in lymphocytes?
   a. Mitotic recombination germline DNA
   b. Junctional flexibility in splicing of exons during mRNA formation
   c. Combinatorial association
   d. All of the above

9. What is the specific site on a foreign substance to which an antibody binds or which activates antibody synthesis?
   a. Epitope
   b. Antigen
   c. Bursal region
   d. Hypervariable site
   e. Domain

10. Passive immunity occurs when
    a. lymphocytes remain fixed and ingest only foreign material striking them.
    b. no response occurs on first exposure to foreign material.
    c. antibody from another individual is provided to destroy a foreign antigen.
    d. a foreign antigen is tolerated.
    e. None of the above

11. Which of the following is a property that is *not* needed for a foreign substance to be immunogenic?
    a. Molecular weight of greater than 20,000 daltons
    b. Foreignness
    c. Hydrophobic
    d. Heterogeneity of chemical composition
    e. Susceptibility to processing and presentation

12. Self-/not self-recognition is important to cancer cell surveillance because
    a. host cell antigens are tolerated.
    b. tumor antigens on cancer cells are recognized as not self and the cells destroyed.
    c. interruption of blood supply blocks their detection.

d. histamine from mast cells destroys memory cells recognizing tumor antigen, resulting in tolerance.

e. All of the above

f. None of the above

13. Immunologic memory means

a. mutation is enhanced.

b. secondary immunologic responses are more rapid and intense.

c. phagocytes move more rapidly after the antigen.

d. antigen learns to avoid the immune system.

e. macrophages carry antigens to the cerebrum and deposit them there.

Short Answer

14. Explain why vaccinations are effective in protecting against particular infections.

15. What is the concept of tolerance?

16. What domains on an antibody molecule lead to the specificity for binding a particular epitope yet allow different lymphocytes to specifically respond to different epitopes?

Essay

17. Briefly describe the formation of each type of lymphocyte.

18. Describe the processes occurring in a specific lymphocyte that results in the formation of mRNA for one of the chains of an antibody.

Chapter Test Answers

1. **True**

2. **True**

3. **False**

4. **False**

5. **True**

6. **d**

7. **f**

8. **d**

9. **a**

10. **c**

11. **c**

12. **b**

13. **b**

14. Vaccinations induce immunologic memory to a foreign organism or toxin by exposing the host to preparations antigenically identical to the foreign substance without the physiologic dysfunction.

15. Tolerance occurs when the host fails to mount an expected immunologic response. Three states seem to induce tolerance: self antigens, low zone where little antigen is present, and high zone where abundant antigen occurs.

16. The variable domains and even more the hypervariable subdomains of both the light and heavy chains have the potential to provide a large number of proteins, each of which can fold in different ways because of the high rate of mutation that occurs in these regions of the DNA coding for these domains.

17. Some stem cells in the bone marrow differentiate into noncompetent virgin lymphocytes that migrate via blood to the thymus gland or bursal regions of bone marrow. Lymphocytes in the thymus gland undergo "processing" to become T lymphocytes and move via blood to the paracortical region of secondary lymphoid organs. Lymphocytes in bursal regions become B lymphocytes and travel via blood to the cortex of secondary lymphoid organs. When activated, T lymphocytes may produce cell-associated antibody, and B lymphocytes will produce humoral antibody.

18. Using heavy chain formation as an example (Figure 14.6), the first step is the mitotic recombination between the V and D regions in the DNA. Next, a second recombination event joins the VD region with a J region on the DNA. On activation of the lymphocyte, a primary RNA transcript is produced from which introns and extra exons must be removed by splicing. Splicing may be accompanied by nucleotide deletion at the P-region or nucleotide addition at the N-region. The resulting mRNA migrates to the cytoplasm where it is translated into a protein.

Check Your Performance

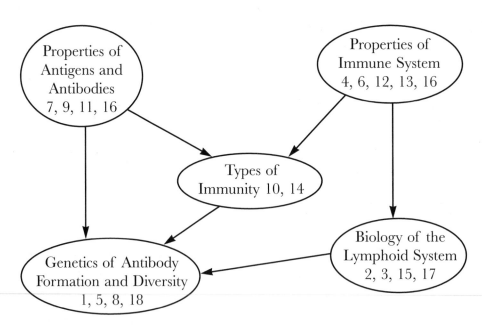

Note the number of questions in each grouping that you got wrong on the chapter test. Identify areas where you need further review and go back to relevant parts of this chapter.

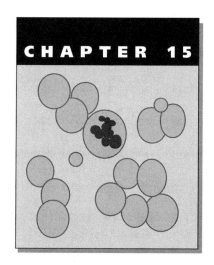

Antigen-Antibody Reactions in Host Defense and in Serology

The immune system, the last line of defense, must respond strongly with devastating effect on the foreign substance because the other two lines of defense have either failed or are stressed. The killing power of this system must be extensive yet focused specifically on the foreign material. Upon removal of the foreign material, the system must be dampened down.

Once virgin immunocompetent B and T lymphocytes have seeded the lymph nodes and spleen, they await activation. Activation requires the interface of three types of cells: **antigen-presenting cells**, such as macrophages, dendritic cells, or B lymphocytes; **T helper (T_H) lymphocytes** to amplify the signal; and **B lymphocytes** to produce humoral antibody or **T cytotoxic (T_C) lymphocytes** to release cytotoxins. The immune system is potentially destructive to the host; hence, it is difficult to activate. However, once activated, the system has positive feedback mechanisms that not only maintain its level of response but also enhance it until the crisis is over. Bringing three different types of cells together requires intercellular communication that is accomplished through cell contact with specific cell surface components, **major histocompatibility complex (MHC)**, and chemical messengers, **cytokines**.

ESSENTIAL BACKGROUND

- Properties of antigens and antibodies (see Chapter 14, Topic 1)
- Biology of the lymphoid system (see Chapter 14, Topic 4)
- Properties of the immune system (see Chapter 14, Topic 2)

TOPIC 1: ACTIVATION AND RESPONSE OF THE IMMUNE SYSTEM

KEY POINTS

✓ *What is the process of presenting an epitope?*

✓ *How do presenting cells, T_H lymphocytes, and B lymphocytes interact?*

✓ *How is an activated immune system dampened down?*

Before the three cell types can develop an immune response, they must be able to recognize each other. Part of the recognition process relies on specific membrane surface components called **major histocompatiblity complex** (**MHC**). MHC I is found on most nucleated cells in an individual and serves in self-recognition but also is important to cells presenting antigens to activate T_C lymphocytes. MHC II is found on surfaces of cells presenting antigens to T_H lymphocytes. Multiple signals ensure the unleashing of the destructive forces of B or T lymphocyte activation. Which B or T lymphocyte clones are activated depends on strong bonding between receptor molecules on B or T lymphocyte surfaces and the MHC-antigen complexes on presenting cells, then secondary binding between other surface molecules on the interacting cells can occur.

Antigen processing by macrophages, dendritic cells, other macrophage-like cells, and B lymphocytes of the secondary lymphoid tissues activates the immune system. All **antigen-processing cells** trap foreign antigens in **phagosomes** (**Figure 15.1**). B lymphocytes incorporate into their cell membranes **B cell receptors** (**BCRs**) having binding sites identical to the binding

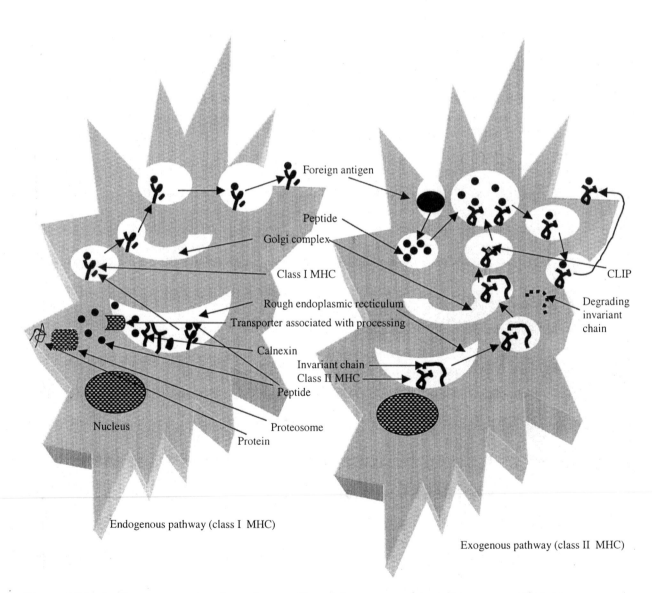

Figure 15.1. Antigen processing in a phagocytic cell to prepare for epitope presenting to lymphocytes.

sites on the humoral antibodies they will produce. Only antigens with a particular epitope bind to the BCRs of a specific B lymphocyte and will be ingested endocytotically by that B lymphocyte. Phagocytic cells are nonspecific in their attraction to and ingestion of antigens. Lysosomes in the cytoplasm of the presenting cells fuse with the phagosome, emptying their degradative enzymes into the phagosome. Antigens are digested partially into an array of different fragments probably between 20,000 and 100,000 daltons having one or more epitopes. Phagosomes containing antigenic fragments fuse with vesicles holding MHC II or MHC I, depending on whether the antigen was exogenous (MHC II) or endogenous (MHC I), such as a virus. In the case of B lymphocytes, antigenic fragments bind to BCRs. The MHC or BCR complexes migrate to the cytoplasmic membrane positioned so that the antigenic fragment projects from the cell surface, making the cell a **presenting** one. If the antigen-presenting cell is located distant from a lymph node or spleen, such as Langerhans cells, they will migrate to the nearest lymph node. In lymph nodes or spleen they will **present** their antigenic fragment to T_H lymphocytes.

When macrophages or dendritic cells present to T_H lymphocytes, three steps must occur before the T_H lymphocyte will activate:

1. MHC II-antigen fragment complex must bind to a T-cell receptor (TCR) specific for the exposed epitope on the antigenic fragment presented. CD 4 helps stabilize this binding.

2. B 7 on the presenting cell must bind to CD 28 on the T_H lymphocyte.

3. Interleukin (IL)-1 from the presenting cell binds to IL-1 receptors on the T_H lymphocyte.

The T_H lymphocyte undergoes mitotic division and produces IL-2, IL-4, and IL-5. Presenting B lymphocytes are drawn to the area of activated T_H lymphocytes, probably by IL-2. Another three-step process must occur to activate the B lymphocytes (Figure 15.1):

1. BCR-antigen fragment complex binds to TCR on the T_H lymphocyte;

2. CD 40 on the B lymphocyte with CD 40L on the T_H lymphocyte;

3. IL-2, IL-4, and IL-5 from T_H lymphocyte bind to respective receptors on the B lymphocyte.

The B lymphocyte begins dividing mitotically, producing some small B lymphocytes that become **memory cells** and a large number of large **plasma cells** that produce and release humoral antibody. Antigens requiring B lymphocytes to have T_H lymphocyte contact are **thymus-dependent antigens**.

If macrophages or dendritic cells are presenting MHC I-antigen fragment complexes, their target cells are T_C lymphocytes. A three-step process is needed for T_C lymphocyte activation (**Figure 15.2**):

1. The MHC I-antigenic fragment complex binds to TCR, and then CD 8 stabilizes the reaction on the T_C lymphocyte;

2. B 7 must contact CD 28;

3. IL-1 from the presenting cell completes the activation of the T_C lymphocyte to divide and produce cytotoxins.

Natural killer (NK) cells, large granular lymphocytic cells having nonspecific cytotoxicity within the lymphoid system, attack tumor or virus-infected cells. MHC I interaction is significant in the response of NK cells to the host cells they contact. Tumor or virus-infected cells have reduced MHC expression. When NK cells contact tumor or virus-infected cells, glycoproteins on

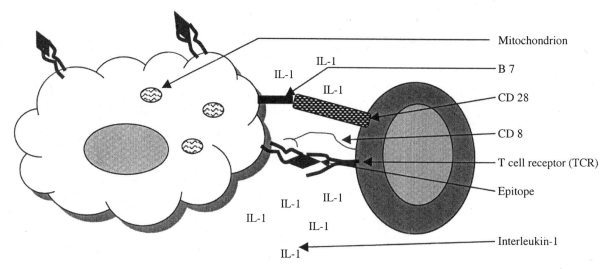

Figure 15.2. Activation of a T$_C$ lymphocyte by a presenting macrophage.

the target cells bind NKR-P1 proteins on the NK cell membrane (**Figure 15.3**). NK cells then release their cytotoxins. When healthy cells contact NK cells, glycoproteins on the normal cells bind NKR-P1 on NK cells, but also MHC I on normal cells react with Ly 49 on NK cells. The second complex reverses or blocks the signal from the first contact site, and NK cells retain their cytotoxins.

Complement, a set of about 30 humoral and membrane proteins, sustains and amplifies the initial antigen-antibody reaction until the foreign substance is removed. Complement can be activated by antibody "tails" of antigen-antibody complexes, the **classic pathway**, or by attachment of complement 3 (C3) protein to microbial surfaces, the **alternative pathway**. Antigen-antibody complexes trigger a cascade of reactions involving humoral complement proteins, enhancing phagocytosis, and sustaining B lymphocyte response (**Figure 15.4**). At the termination of the complement cascade reactions, a **membrane-attack complex** (**MAC**) is formed. MACs attach to cellular membranes, both foreign and "**innocent bystanders**," displacing phospholipids to form large transmembrane channels that permit ions and small molecules to diffuse through the membranes freely.

When foreign substances have been removed, the immune system must be dampened to prevent destruction of the host. Although the exact mechanisms for reducing immune responses are not known, the process is prolonged. Prolonging the shutdown, that is, dampening, the immune system ensures no "hidden pockets" of foreign antigen remain in the host and also ensures that if reinfection should occur shortly after clearing, a strong immune response can be mounted quickly. Two mechanisms have been proposed: **suppressor cells** and **idiotype networks**. Certain T lymphocytes (**T$_S$**) appear to release cytotoxins that block T$_H$ and B lymphocyte responses. Another class of lymphocytes found in bone marrow and having large granular inclusions secretes **suppressor factors** (**SFα** and **SFβ**) that suppress B and T lymphocyte proliferation and immunoglobulin activity. The other mechanism involves antibodies directed against other antibodies. When antibodies bind antigens, conformational changes in the antibody occur in the antigen-binding domain. These changes create new epitopes (**idiotopes**) on the antigen-antibody complex that are considered not self by the immune system. The antigen-antibody complex activates a B lymphocyte clone to produce antibody against the original antibody, in

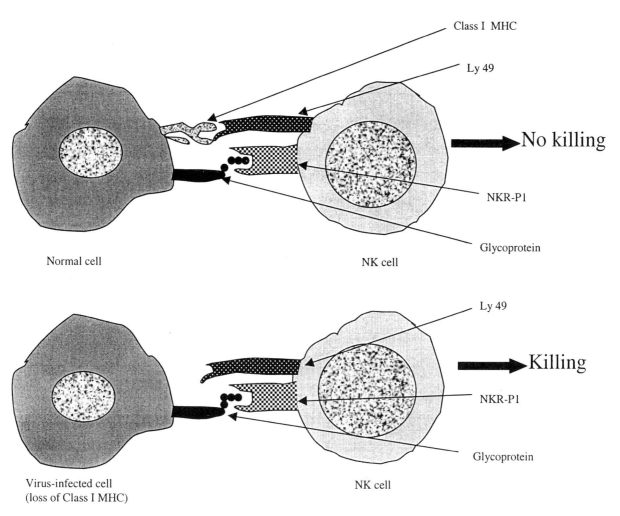

Class I MHC

Ly 49

No killing

NKR-P1

Glycoprotein

Normal cell

NK cell

Ly 49

Killing

NKR-P1

Glycoprotein

Virus-infected cell
(loss of Class I MHC)

NK cell

Figure 15.3. Activation of NK cells by contact of NKR-P1 and glycoproteins on a virus-infected cell, resulting in killing of the infected cell.

other words, an anti-antibody. This second antibody in turn generates new epitopes when it binds to its antibody target. Other B lymphocyte clones are stimulated to produce anti-anti-antibodies. Theoretically, this process could proceed ad infinitum. From a practical standpoint, probably the process goes only three to four layers deep, nonetheless forming an **idiotype network**. Dilution of later idiotypes by phagocytosis of antigen-antibody complexes undoubtedly restricts extension of the idiotype network.

Topic Test 1: Activation and Response of the Immune System

True/False

1. Activation of the immune system requires only the contact of a macrophage with a B lymphocyte.

2. Use of two-step contacts between presenting cells and responding cells ensures that the immune system is not casually activated.

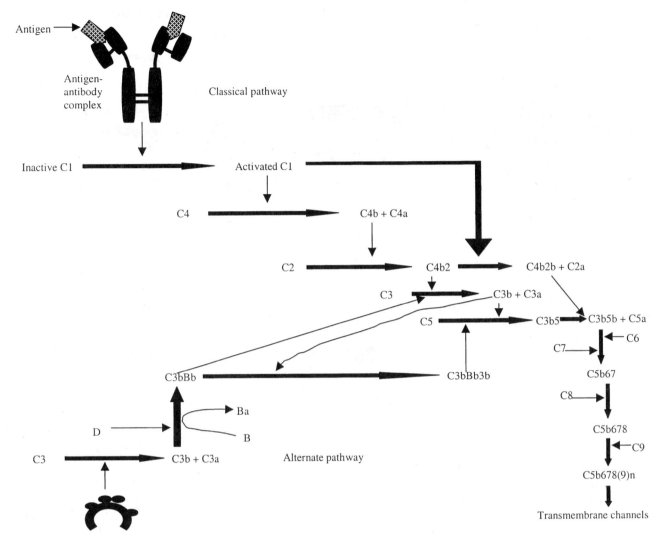

Figure 15.4. Activation and cascade of reactions involving complement proteins in serum.

Multiple Choice

3. Which of the following substances is *not* involved in activation of T_H lymphocytes?
   a. NKR-P1
   b. B 7
   c. MHC II-antigen fragment complex
   d. IL-1
   e. None of the above

4. Which of the following describes the effect of complement proteins on B lymphocytes?
   a. Activate T_H lymphocytes
   b. Antibody synthesis of B lymphocytes enhanced
   c. Suppress mitosis of B lymphocytes
   d. Inhibit macrophage presenting
   e. Attach membrane attack complexes to B lymphocyte membranes

Short Answer

5. What mechanisms may dampen immune responses?

Topic Test 1: Answers

1. **False.** Macrophages and dendritic cells present processed antigen fragments on MHC II to T_H lymphocytes. Then T_H lymphocytes contact presenting B lymphocytes that become activated.

2. **True.** The initial contact between MHC II-antigen fragment complex must be tight with TCR and CD4 on the T_H lymphocyte. Quickly thereafter, B 7 and CD 28 must bind.

3. **a.** NKR-P1 is found on the surface of natural killer cells and is involved in tumor or virus-infected cell destruction.

4. **b.** B lymphocytes have CR-1 and CR-2 receptors that bind C3a coated antigen-antibody complexes.

5. T_S lymphocytes or natural suppressor substances may contact T_H or B lymphocytes and release cytotoxic compounds. Alternatively, idiotypic networks of anti-antibodies to new epitopes (idiotopes) formed when antigen-antibody complexes occur. In some way these networks dampen B cell release of antibody.

TOPIC 2: HYPERSENSITIVITIES: TYPES I, II, III, AND IV

KEY POINTS

✓ *What distinguishes each type of hypersensitivity?*

✓ *What is an allergen?*

✓ *How does each type of hypersensitivity affect a host?*

During intrauterine life, probably shortly before birth or hatching, an individual's immune system determines every potential antigen present within the individual will be a **self antigen** and tolerated. All other antigenic materials will be considered **not self**. Not all foreign antigens are harmful to an individual, for example, milk proteins or wheat glutin. Because these are foreign antigens, an immune response may be instituted. An immune reaction to harmless antigens is considered inappropriate and is called an **allergy**. The inducing antigen is an **allergen**. Haptens can become foreign epitopes if they bind to host proteins or lipopolysaccharides, for example, the antibiotic penicillin. These new epitopes, being not self, will become allergens. An allergen can induce one of four allergenic responses (**Table 15.1**). Some allergens elicit different allergies in different people and can trigger different allergies in the same individual on different exposures. The physiologic responses of allergies are not expressed upon the initial exposure to an allergen. Allergies are secondary immune responses.

Table 15.1. Different Types of Allergic Reactions

Type of Allergy	Antibody Involved	Affected Agents	Physiologic Response	Examples
Anaphylaxis (type I)	IgE	Mast cells	Histamine released	Hay fever, common cold
Antibody-dependent cell-mediated cytotoxicity (type II)	IgG	Usually erythrocytes coated with antigen or complement fixation	Hemolytic anemia	Transfusion reactions, erythroblastosis fetalis
Arthus reaction (type III)	IgG	Complement fixation	Complement activation of mast cells, histamine released	Glomerulonephritis, rheumatoid arthritis
Delayed hypersensitivity (type IV)	T-cell bound antibody	T_{DHT} lymphocytes	Cytotoxins released	Poison ivy, drug allergies

Topic Test 2: Hypersensitivities: Types I, II, III, and IV

True/False

1. All hypersensitivities can arise on the initial exposure of the allergen.

2. Allergies are inopportune immune responses.

Multiple Choice

3. Anaphylactic responses occur when
   a. complement protein fragments activate macrophages.
   b. activated lymphocytes T_{DHT} lymphocytes release INFγ, MIF, and TNFβ.
   c. allergen-IgE antibody complexes bind to mast cells, releasing histamine.
   d. allergen-IgG antibody complexes precipitate toxin release from phagocytic cells.
   e. All of the above

4. What may be activated in ADCC (antibody-dependent cell-mediated cytotoxicity) hypersensitivity?
   a. IL-2
   b. Complement
   c. IL-1
   d. Superoxide
   e. Fibronectin

Topic Test 2: Answers

1. **False.** Allergies arise from secondary immune responses as a result of immunologic memory to epitopes found on the allergen or to new epitopes formed because the allergen reacts with host proteins.

2. **True.** Allergens are rarely harmful to the host, but they are foreign immunogens or form foreign immunogens with host proteins.

3. **c.** Allergen-IgE complexes bind to IgE tail receptors on mast cells, resulting in the emptying of histamine-containing vesicles outside the mast cell.

4. **b.** IgG-antibody complexes bind to complement proteins leading to MAC complexes on target cells.

TOPIC 3: AUTOIMMUNE DISEASES

KEY POINTS

✓ *Why would the body destroy itself with its own immune system?*

Autoimmunity arises when an antigen that should be considered **self** is deemed **not self** and an immune response is mounted. The effect on the individual will depend on organ-specific autoimmunity or systemic autoimmunity. Which cells of an organ are targeted also determines the physiologic state of the individual. Four mechanisms have been proposed to explain the induction of autoimmunity: release of sequestered antigens, molecular mimicry of a foreign antigen, inappropriate expression of MHC II molecules, and polyclonal B lymphocyte activation (**Table 15.2**). Because of the wide range of tissues affected, it is unlikely a single mechanism is involved.

Topic Test 3: Autoimmune Diseases

True/False

1. Autoimmunity occurs when an individual's immune system mounts an attack against one of that individual's proteins.

2. Autoimmunity of proteins of the myelin sheath leads to multiple sclerosis.

Table 15.2. Examples of Autoimmune Diseases				
DISEASE	TISSUE INVOLVED	SELF-ANTIGEN TARGETED	PHYSIOLOGIC RESPONSE	POSSIBLE MECHANISM
Hashimoto's thyroiditis	Thyroid gland	Thyroglobulin, microsomal proteins	Thyroxin insufficiency	Release of sequestered antigen
Pernicious anemia	Intestinal membranes	Intrinsic factor	Loss of iron uptake iron deficiency	Molecular mimicry
Insulin-dependent diabetes mellitus	Islets of Langerhans pancreas	Beta cells	Insulin deficiency	Release of sequestered antigen, inappropriate MHC II molecules
Systemic lupus erythematosus	Many tissues, particularly erythrocytes, leukocytes, platelets	DNA, histones, clotting factors	Flushing of body surface, light sensitivity	Polyclonal B lymphocyte activation
Multiple sclerosis	Schwann cells around neurons	Myelin sheath	Progressive neurologic dysfunction	Release of sequestered antigen?
Myasthenia gravis	Muscle	Motor end plates	Progressive muscle weakening	Molecular mimicry

Multiple Choice

3. In myasthenia gravis, autoantibodies
   a. inactivate motor end plates.
   b. are cross-reactive between bacterial heat shock proteins and host heat shock proteins.
   c. trigger T_{DHT} lymphocyte destruction of beta cells in pancreatic islets of Langerhans.
   d. block uptake of iodine needed for thyroxin synthesis.
   e. activate T_C lymphocytes to attack synovial membranes.

Topic Test 3: Answers

1. **True.** This is the definition of autoimmunity.

2. **True.** In multiple sclerosis myelin sheath is attacked, destroying the insulation around axons of nerves and leading to short-circuiting of the electrochemical signals in the neurons.

3. **a.** Because acetylcholine is not bound to membrane receptors, muscle cells are not activated and do not contract.

TOPIC 4: MONOCLONAL ANTIBODY PRODUCTION BY HYBRIDOMA CELLS

KEY POINTS

✓ *What is a hybridoma?*

✓ *How is a single type of antibody isolated when most antigens have multiple epitopes?*

Identification, localization, and separation of a particular protein in the cellular milieu have been important for our current understanding of the cell. The specificity of an antibody for a specific epitope provides a method for targeting particular proteins in a milieu. To ensure specificity, one must be certain the antibody comes from one type of lymphocyte, a **monoclonal antibody**.

One method of getting lymphoid cells that can be cloned and produce only one type of antibody is to form a **hybridoma** (**Figure 15.5**). A **hybridoma** is the fusion of two cell types, one of which is immortal. Myeloma tumor cells were chosen because they are of lymphoid origin, are immortal, and mutants unable to salvage purines (hypoxanthine guanine phosphoribosyl transferase negative, HGPRT⁻) or to synthesize immunoglobulin (Ig⁻) can be isolated. Once isolated, one can clone these cells indefinitely in tissue culture.

The second cell in the hybridoma is an activated B lymphocyte from mouse spleen. The mouse is injected with the antigen of interest before harvesting the spleen cells. The B lymphocytes produce antibody and have HGPRT.

Cells fused with polyethylene glycol are recognized microscopically because they have two nuclei. Cells are cultured in HAT medium (*h*ypoxanthine, *a*minopterin, *t*hymine) in well plates. Aminopterin blocks de novo synthesis of nucleotides, and the HGPRT mutation blocks salvage pathway of nucleotide synthesis in mutant myeloma cells. Highly differentiated activated B lymphocytes suffer apoptosis after a few subdivisions. Only hybridoma cells are able to multiply in HAT medium. A few wells will contain small clusters of cells after 7 to 10 days. Each cluster

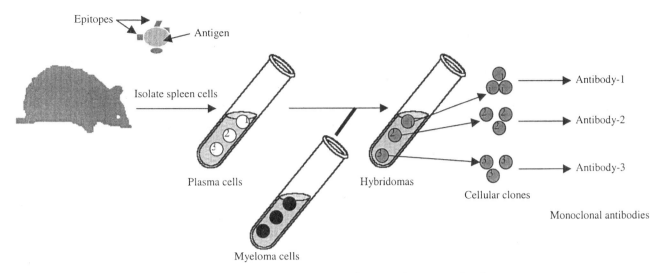

Figure 15.5. Constructing hybridomas for the synthesis of monoclonal antibodies.

is the clonal expansion of a hybridoma cell. Single cells from a cluster are isolated in separate wells of HAT medium and subcloned to ensure that monoclonal antibody will be produced.

Aliquots of the cell-free medium from hybridoma subcultures are assayed for the desired antibody. Usually serologic techniques, such as enzyme-linked immunosorbent assay (ELISA) and radioimmunoassay (RIA), are used to detect the antigen-antibody reaction (see Topic 5). Once a hybridoma is found that produces the desired monoclonal antibody, those hybridoma cells are freed from the plastic wells and subcloned in HAT medium in large tissue-culture flasks or in the peritoneal cavity of histocompatible mice. Hybridomas cultured in the peritoneal cavity of mice release 10 to 100 times more monoclonal antibody into ascites fluid than hybridomas cultured in tissue culture produce. After collection of tissue culture medium or ascites fluid, the monoclonal antibodies are purified by column chromatography.

Topic Test 4: Monoclonal Antibody Production by Hybridoma Cells

True/False

1. Antibodies collected from a serum sample are monoclonal.

2. B lymphocytes from spleen are immortal because they are dividing.

Multiple Choice

3. To eliminate myeloma cells that did not hybridize to form hybridomas, what must one do?
   a. Make sure myeloma cells are immunoglobulin synthesizers.
   b. Starve them of polyunsaturated fats.
   c. Select HGPRT lacking cells and culture them in aminopterin.
   d. Add T_H lymphocytes to the cell culture.
   e. Put 0.05% sodium hypochlorite in the medium.

4. B lymphocytes are fused to myeloma cells using what material?
   a. Polysorbitan
   b. Collagen
   c. Triton X-100
   d. Polyethylene glycol
   e. Isopropanol

Short Answer

5. What is done to ensure monoclonal antibody is being synthesized?

Topic Test 4: Answers

1. **False.** When serum is collected, it contains antibodies produced against several epitopes by different lymphocyte clones. No separation of clones is done.

2. **False.** Although B lymphocytes do divide for several generations, they are highly differentiated and soon experience apoptosis.

3. **c.** Cells that are HGPRT negative cannot use the salvage pathway to synthesize purine nucleotides, and aminopterin in the medium blocks de novo nucleotide synthesis. Hence, the myeloma cells die because they cannot synthesize the nucleotides needed for DNA synthesis.

4. **d.** Polyethylene glycol induces fusion of cellular membranes.

5. When clusters of hybridoma cells are first cultured, the small clusters are broken and individual hybridoma cells are placed in HAT medium in separate wells in a well plate.

TOPIC 5: SEROLOGY

KEY POINTS

✓ *How can antigen-antibody complexes be recognized?*

✓ *When is each technique best used?*

If one is to understand how the immune system works, it is essential to measure when antibody is present and how much has been produced. **Serology** is the discipline devoted to the development of techniques to measure antigen-antibody reaction and the use of those techniques. **Antibody affinity** is a measure of the strength of the total noncovalent interactions between an antigen binding site on the antibody and an epitope on the antigen and significant for the detection of **cross-reactivity**. **Cross-reactivity** occurs when antigens from different sources have common or similar epitopes. The affinity is usually stronger for the antigen that induced antibody synthesis. For example, the vaccine used to limit variola virus (smallpox) is active against generally avirulent vaccinia virus (cowpox). In another example, ABO surface antigens on erythrocytes have epitopes that bind antibody directed against common intestinal bacteria. Some of these assays are quantitative, whereas others are qualitative and only intended to detect antigen-antibody complexes. The sensitivity of the various assays varies.

Radioisotopes can be used to tag either antigen or antibody. Detection or monitoring of drugs in circulation can be achieved quantitatively by using a **competitive RIA**. Patient's serum containing unlabeled drug is mixed with an excess of anti-drug antibody, and then a known quantity of radiolabeled drug is added. After precipitation of antigen-antibody complexes, radioactivity in the supernate is measured. Unlabeled drug in the patient's serum bound to the antibody, forcing some of the added labeled drug to remain in solution. The amount of radioactive drug in solution equals the amount of unlabeled drug in the patient's serum. **Solid-phase RIA** can be used to detect the presence of a viral pathogen in a patient's serum. Here, microtiter wells are coated with a constant amount of antibody to a surface antigen on the virus. A sample of the patient's serum and ^{125}I-labeled antigen are added and allowed to bind to the antibody. Supernatant is removed and the amount of radioactivity bound to the antibody is determined. If the serum sample contains virus, the amount of labeled antigen bound will be less than in controls with uninfected serum samples. Quantification is achieved using a standard curve constructed with varying amounts of labeled antigen. To detect qualitatively small amounts of a protein in a mixture, one could use **autoradiography** with a ^{125}I-labeled monoclonal antibody specific for the protein. A protein mixture would be electrophoresed in a gel, such as polyacrylamide gel, and then a replica made by blotting the proteins from the gel onto a filter. The filter is submersed in radiolabeled monoclonal antibody that binds only to the protein band that is antigenic. X-ray film is placed over the filter. After exposure, when the film is developed, a dark band appears wherever the radioactive antibody was bound to the filter by the antigenic protein.

Enzyme-tagged antibodies are used in **ELISA**. Indirect ELISA is the major method for detecting if a blood unit is free of human immunodeficiency virus (HIV) (**Figure 15.6**). Because circulating free virus is minimal usually, one must search for anti-HIV antibody in the blood. HIV viruses are bound to the plastic edges of the wells. The wells are washed and the serum to be tested is added to the wells. If anti-HIV antibodies are present, they will bind to the HIV virus. The wells are washed, and goat or rabbit anti-human serum antiserum tagged with an enzyme, such as horseradish peroxidase, is put in the wells. If known anti-HIV antibodies are present and bound to the HIV, then the enzyme-tagged anti-human serum antiserum binds to the human anti-HIV antibodies. The wells are washed to flush away any unbound enzyme-tagged antibody. Next, enzyme substrates are added and a colored product will form if any enzyme-tagged antibody was bound in the well. The colored products indicate anti-HIV antibody in the blood

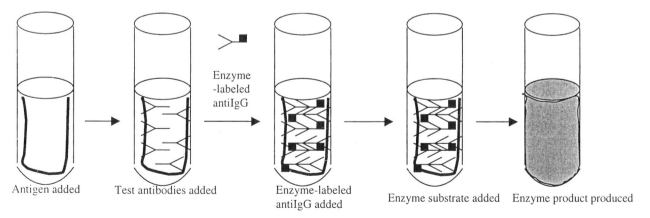

Figure 15.6. Indirect enzyme-linked immunosorbent assay (ELISA), such as that used to detect HIV antibody in a patient's serum.

sample, indicating HIV infection. The intensity of the color can be used to quantitate the amount of anti-HIV antibodies bound.

Fluorescent-tagged monoclonal antibodies tagged with fluorescein and rhodamine can detect antigens in situ in tissue specimens or in cell cultures placed on slides. An indirect test, **fluorescent treponemal antibody absorption (FTA-ABS) test**, confirms the presence of antibodies to *Treponema pallidum* in individuals suspected of tertiary stage syphilis (see Chapter 16, Topic 3). Antigens can be detected in situ in specimens prepared for electron microscopy using ferritin-labeled antibody. The dense iron atoms in the ferritin scatter the electron beams, and dark areas on electron micrographs indicate where in the cell the antigen is located.

Soluble antigens can be detected by **precipitation** of the clumped antigen-antibody complexes. If the antigen is particulate, **agglutination** (clumping) of the particles indicates antigen-antibody complex formation. **Precipitin** reactions are dependent on the ratio of the antigen concentration to antibody concentration, usually two antibody molecules to three antigen molecules gives large enough antigen-antibody complexes to have precipitation. Lower concentrations or higher concentrations of antigen prevent the formation of a precipitate.

Immunodiffusion techniques are used when one expects mixtures of either antigens or antibodies and some separation is needed before permitting antigens and antibodies to complex. These assays detect cross-reacting antigens and are usually qualitative. In the **Ouchterlony double diffusion** technique, wells are cut around the periphery of an agarose gel in a Petri plate. A well is cut in the center of the gel. The contents put in the center well depend on whether one is looking for a particular antigen in specimens or whether one is determining if an antibody is present in serum specimens. In the former case, antibody to the antigen is placed in the center; in the latter case, antigen goes in the center well. Diffusion occurs and precipitation bands are formed between the wells when antigen-antibody ratios are favorable. With **immunoelectrophoresis** through agarose gels, a protein mixture is separated first by using an electrical field (**Figure 15.7**). Proteins are separated by their charge and their molecular weights. When the electrical field is removed, polyclonal antibody is placed in a trough parallel to the electrophoresis lanes and diffusion occurs. Curved bands of precipitin occur where antigen-antibody complexes form.

Agglutination reactions use particles either naturally coated with antigen, such as erythrocytes, or ones artificially coated with soluble antigens, for example, erythrocytes or latex beads. The particles are mixed with sample sera. If specific antibody is present, the particles clump tightly. By making serial dilutions of the antibody specimens, a semiquantitative assay can be done, looking for the lowest dilution at which clumping still occurs.

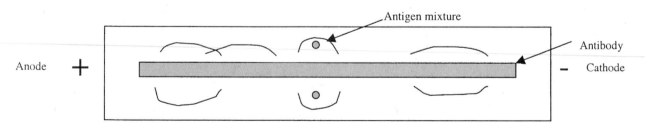

Figure 15.7. Precipitin bands observed after immunoelectrophoresis.

Complement fixation involves pitting two antigen-antibody reactions against one another. The known reaction involves binding sheep erythrocytes to hemolysin (rabbit anti-sheep erythrocyte antiserum). The experimental reaction concerns testing for the presence of an antigen in some material or of antibody to the antigen in question in serum (sera). The competition is for binding guinea pig complement that is very sensitive in this assay. First, the experimental reaction complexing the antigen and antisera in question is performed, and then guinea pig complement is added. If there is an antigen-antibody reaction, complement will be bound to this complex; otherwise complement will remain free. Next, the hemolysin and sheep erythrocytes are complexed and then added to the complement-containing mixture. If there is free complement, it will bind to the hemolysin-tagged erythrocytes, and hemolysis will occur. This assay can be semiquantitative if serial dilutions of either the experimental antibody solutions or experimental antigen solutions are made before starting the series of complement fixation reactions. This assay has a clear detection point but requires standardization of reagents and many controls to ensure that hemolysis is the result of complement binding to the hemolysin-complexed sheep erythrocytes. In this test, hemolysis indicates the absence of either the experimental antigen or the experimental antibody under consideration.

Topic Test 5: Serology

True/False

1. All techniques discussed for measuring antigen-antibody complexes require using monoclonal antibodies.

2. Soluble antigens can be detected when they precipitate after binding antibody.

Multiple Choice

3. Which of the following is *not* used to tag an antigen or an antibody?
   a. Ferritin
   b. Radioactive ^{125}I
   c. Fluorescein
   d. Horseradish peroxidase
   e. None of the above

4. Antigen-antibody complexes are detected in immunoelectrophoresis and Ouchterlony double diffusion by
   a. precipitin bands formed.
   b. fixation of complement to erythrocytes.
   c. fluorescent bands.
   d. Radioactivity.
   e. product deposition.

Short Answer

5. Which of these techniques would be useful for locating a protein in a cell?

6. Why is there said to be competing reactions in the complement fixation technique?

Topic Test 5: Answers

1. **False.** In some, such as immunoelectrophoresis, one frequently wants to detect several antigens and uses polyclonal antibodies. Also in some tests, such as HIV infection, one is seeking any type of antibody that has been synthesized against any of the epitopes on the antigen.

2. **True.** As long as the proper ratio of antigen to antibody occurs, the complex will precipitate.

3. **e.** All the above choices are used to tag either antibody or antigen molecules.

4. **a.** In both cases, antibody and antigen molecules diffuse through agarose gels until they meet in the right ratios for precipitation.

5. RIAs followed by autoradiography, fluorescent-labeled antibody, or ELISA are able to bind to a protein in situ and be detected without destroying the cells.

6. Because complement is introduced into the solution in which the experimental antigen-antibody complex is supposed to have formed. Then, erythrocyte-hemolysin complexes are added. If there is any free complement after exposure to the experimental complexes, that complement will bind to the hemolysin bound erythrocytes, inducing hemolysis.

APPLICATION

Suppose we are in west Texas where cattle and oil reign supreme on the arid plains. A rancher has been experiencing losses from his grazing cattle herds because cattle are being killed and butchered in his isolated pastures. A truck has been seen in the county seat with blood on its bed-liner. When stopped and questioned, the truck's owner claimed he had carried two injured hunting dogs to the vet. Nonetheless, because this truck had been observed near the rancher's fields, warrants were obtained to get samples of the blood in the truck. An Ouchterlony double diffusion experiment was initiated. Reconstituted serum samples from the truck bed were placed in the center well on the agarose gel plate. Antisera to several animals, including dogs and cattle, were placed in the peripheral wells. After incubation, precipitin bands were observed between anti-dog serum antiserum and the central well, but none anywhere else on the plate. Also, aliquots of the erythrocytes recovered from the truck bed were mixed individually with several antisera against the erythrocytes of various animals, including dogs and cattle. Agglutination of erythrocytes was sought. Only the anti-dog erythrocyte antiserum induced agglutination of the recovered erythrocytes. On the basis of this evidence, the truck owner was exonerated of any crime.

Chapter Test

True/False

1. CD refers to clone designation to indicate which clone of lymphocytes was stimulated to produce antibody.

2. Hybridomas are formed from the fusion of cells from two different clones of lymphocytes.

3. When using the indirect FTA-ABS test, the surface of *Treponema pallidum* cells will fluoresce in ultraviolet light if the patient's serum contains anti-treponemal antibody.

4. The immune system is deactivated abruptly as soon as the foreign antigen is eliminated from the host.

5. Serology is the study of antibody formation in the serum of mammals and birds.

Multiple Choice

6. What happens when a person shows autoimmunity?
   a. B lymphocytes begin to multiply uncontrollably.
   b. Individual synthesizes antibody against his or her own antigens.
   c. Complement spontaneously fixes to the individual's erythrocytes.
   d. Mast cells release histamine.
   e. Macrophages release strong oxidizing compounds.

7. Which of the following serologic tests involves competing antigen-antibody reactions?
   a. Complement fixation
   b. Immunoelectrophoresis
   c. ELISA
   d. Agglutination
   e. Fluorescent-tagged immunoglobulins

8. Systemic type I hypersensitivity may induce extremely serious effects on the patient because
   a. overactive T_C lymphocytes are sacrificed and the person suffers immunodeficiency.
   b. T_H lymphocytes are inactivated, preventing B lymphocyte activation.
   c. inability of macrophages to present "processed" antigenic fragments.
   d. massive release of histamine from antigen-IgE antibody complexes on mast cells, resulting in respiratory arrest.
   e. None of the above

9. What cluster of things must occur to activate a T_H lymphocyte?
   a. Macrophage MHC II-epitope complex binding to TCR, B 7 complexing to CD 28, and IL-1 release from macrophages
   b. Macrophage MHC I-epitope complex binding to BCR, B 7 binding MHC II, and release of CD 43
   c. Binding of C3b to C5a after fragmentation of C4 into C4c and C4d
   d. Binding of polysaccharide to several TCR molecules, then binding of CD4 and CD 28, followed by release of IL-2, -4, -5, and -12
   e. None of the above

10. What is the reaction called that occurs 2 to 3 days after exposure to an antigen and activating T_{DHT} lymphocytes?
    a. Type I hypersensitivity
    b. Cross-reactivity
    c. Autoimmunity

d. Immunodeficiency

e. Type IV hypersensitivity

11. Which of the following serologic tests *cannot* be performed relatively easily quantitatively or semiquantitatively?

a. ELISA

b. Precipitation or agglutination

c. Complement fixation

d. RIA

e. Ouchterlony double diffusion

12. Which of the following cells is *not* involved in an immune response?

a. B lymphocyte

b. Macrophage

c. T_H lymphocyte

d. Neutrophil

e. Dendritic cell

13. When an antibody binds to a hapten that is not on the antigen triggering that antibody's synthesis, it is said what has occurred?

a. Hypersensitivity

b. Cross-reactivity

c. Antibody affinity

d. Autoradiography

e. Agglutination

Short Answer

14. What does each of the cells fused together to form a hybridoma contribute to monoclonal antibody production?

15. Why is it essential to have three signals to activate T_H or B lymphocytes?

16. What mechanisms have been suggested to explain the occurrence of autoimmunity?

Essay

17. Give two theories for dampening down an immune response?

Chapter Test Answers

1. **False**

2. **False**

3. **True**

4. **False**

5. **False**

6. **b**

7. **a**

8. **d**

9. **a**

10. **e**

11. **e**

12. **d**

13. **b**

14. The B lymphocyte contributes the genetic information for synthesizing a single type of antibody and the myeloma cell gives the hybridoma the capacity for continual cell division.

15. Because the immune system has very strong and potentially destructive effects, the system should not be activated unless it is needed. Use of three signals ensures against inadvertent activation.

16. Four mechanisms have some experimental evidence to explain autoimmunity: release of sequestered antigens, molecular mimicry of a foreign protein, inappropriate expression of MHC II molecules, and polyclonal B lymphocyte activation. There ultimately may be shown that more than one mechanism is involved in initiating autoimmunity.

17. First is the activation of T_S lymphocytes that supposedly block T_H and B lymphocyte division and activity. Second is the idiotype network in which antibodies are directed against idiotypes formed during antigen-antibody binding. Then antibodies are made against the second antibody idiotypes, and so forth.

Check Your Performance

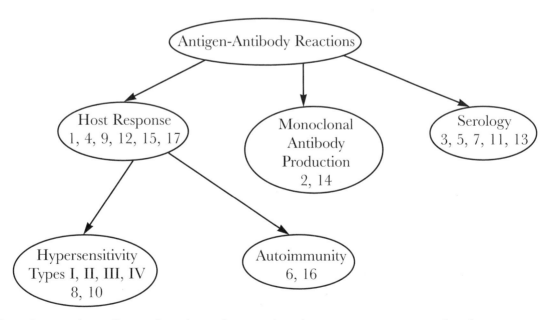

Note the number of questions in each grouping that you got wrong on the chapter test. Identify areas where you need further review and go back to relevant parts of this chapter.

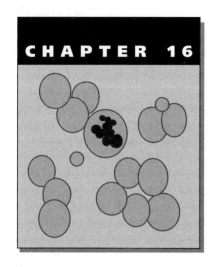

CHAPTER 16

Infectious Diseases Caused by Bacteria

The study of infectious diseases concentrates on the symbiotic relationship of **parasitism**, where one organism benefits at the expense of the other. Parasites are usually much smaller than their hosts. Fortunately for macroscopic hosts, disease is a rare state, the exception and not the rule. This results because the hosts raise efficient defenses against parasites (Chapters 13–15). Pathogens are a subgroup of parasites that induce physiologic dysfunctions in the hosts. Some parasitic microorganisms are opportunistic and become pathogenic when they have the *opportunity*, such as when the host experiences physiologic or psychological stress. Two examples of opportunistic microorganisms are *Streptococcus pneumoniae*, found in the throat of some people, and *Staphylococcus epidermidis*, found on body surfaces of almost every healthy human.

ESSENTIAL BACKGROUND

- **Harmful and beneficial actions of microorganisms in the environment (see Chapter 19, Topic 3)**
- **Symbiotic associations of organisms (Chapter 12)**
- **Physical and chemical agents used to control microorganisms (Chapter 8, Topics 3 and 4)**

TOPIC 1:   BACTERIAL PATHOGENS TRANSMITTED PRIMARILY BY THE AIRBORNE ROUTE

KEY POINTS

✓ *How do these bacteria survive the harsh conditions during transit in air?*

✓ *What portals of entry into a host do these bacteria use?*

Airborne transmission is definitely a stressful experience for a bacterium. Air lacks sufficient nutrients, if it has any. Air is desiccating. How hazardous a journey this is for a bacterium depends on how long it must exist away from the host, if it has characteristics to prevent drying, and if it carries enough stored nutrients to last its airborne existence.

Many bacteria accomplish their airborne journeys in **droplet nuclei** created by the host's coughing, sneezing, or vocalization. Coughing and sneezing are particularly important because the droplets of 10 μm diameter travel at 100 m/s (200 miles/hour), potentially minimizing the period away from a host. This situation is particularly true in crowded or enclosed areas. In addition, the droplet provides moisture, lessening the impact of desiccation.

Dust can be a vehicle for bacterial transmission. Although dust may be very drying and experiences temperature fluctuations, it is spread easily great distances by wind or host activity. The bacteria may spend a prolonged time on the dust particle before reaching a host. Dust particles may aid the bacterium gain a **portal of entry** by physically breaching the outer host defenses (as in scrapes) or by paralyzing host defenses (as with cytotoxic agents). Dust collection in hospitals is unwanted because **nosocomial infections** can arise when dust reaches **immunocompromised** patients.

Diphtheria is a disease of the throat resulting from the poisoning of the epithelial lining of the back of the throat. The lining sloughs off as a grayish gummy **pseudomembrane,** eventually blocking the glottis and leading to asphyxiation. Pathogenic strains of *Corynebacterium diphtheriae* carry **prophage B** that codes for a **toxin** affecting cells of the heart, kidney, and nervous tissues by inhibiting protein synthesis. Vaccination with a nontoxic form of the toxin, a **toxoid**, is the primary method of controlling diphtheria.

Strep throat is one of a number of distinct syndromes started by *Streptococcus pyogenes* found in the throat microflora of some humans. "Flesh-eating" strains are very invasive, producing very rapidly spreading wounds by secreting a battery of cytotoxic compounds. *S. pyogenes* induces toxic shock-like syndrome, scarlet fever, and middle ear infections (otitis), principally in children with small eustachian tubes. Hypersensitivity reactions to *S. pyogenes* components can lead to inflammations of glomerulonephritis in the kidney and rheumatic fever in the heart. Opportunistic invasion by *S. pyogenes* usually initiates strep throat in physiologically or psychologically stressed individuals. Spread is by vocalization, coughing, or improper handling of contaminated materials, such as tissues, throat swabs, and other sample collection tools. Treatment is with penicillin or more potent antibiotics.

S. pneumoniae normally found in the throat can enter the lungs, inducing pneumonia. Normally, the few streptococci entering the lungs are quickly removed by phagocytosis of macrophages. Other *S. pneumoniae* are trapped by mucus and pushed to the throat by action of ciliated epithelial cells of the trachea and bronchi. Anything damaging the ciliated epithelium, such as viruses, toxins, smoking, or corrosive chemical vapors, predisposes the lungs to streptococcal penetration. Many strains of *S. pneumoniae* produce an **abundant capsule** that blocks phagocytosis. *S. pneumoniae* may produce a toxin, **pneumolysin**, destroying cells of the alveolar lining and triggering fluid accumulation and inflammation. Without quick response the patient will die shortly of respiratory collapse. Two other "normal" organisms of the throat or nasopharynx, *Klebsiella pneumoniae* and *Haemophilus influenzae*, may cause similar pathologic symptoms. Culture of phlegm from the lungs is needed to determine which organism is inducing disease because different antibiotics are needed to control these organisms.

Tuberculosis caused by *Mycobacterium tuberculosis* is not very invasive, but these organisms are able to establish themselves within the host, slowly enlarging their niche over years whenever the host faces certain stresses. *M. tuberculosis* has waxy mycolic acids that protect it against dehydration and killing when phagocytized by macrophages. Persistent coughing of infected individuals spreads the organism. Outside a host, this organism has nutrients and a low metabolism to survive prolonged periods of time on surfaces, such as clothing, tabletops, bedding, and eating utensils. When dry, the propagules of *M. tuberculosis* are readily airborne. Persons with normal immune responses will phagocytize *M. tuberculosis* in the lungs but may not kill it. In infants lacking mature immune systems or immunocompromised patients, *M. tuberculosis* can enter the blood and establish progressive infections of one or more organs, leading in a short time to collapse in function of one or more of those organs. Because mycolic acids on the surface of *M. tuberculosis* block delivery of antibiotics to the microorganism, prolonged antibiotic therapy is necessary. Recently, control of tuberculosis has become very difficult because of the appearance of multidrug-resistant *M. tuberculosis*.

Human activities may unwittingly make conditions more favorable for microorganisms to survive in new habitats. These microorganisms may occur commonly in the environment, such as soil, but have had restricted populations because they have fastidious nutritional requirements. The new artificial habitats may be very favorable for these microorganisms. *Legionella pneumophila* requires high iron levels and is **psychrotrophic** (cold tolerant). *L. pneumophila* contaminates evaporative water coolers of large air-conditioning units containing iron structural elements. Infection in the lungs occurs when *L. pneumophila* is blown throughout a building by its air-conditioning unit. No known human to human transmission is known. Other devices with iron piping and chilled water provide sites for *L. pneumophila* populations to develop and spread if contaminated water is sprayed. The best tactic against this disease is to clean the water-containing devices and chlorinate the stored water.

Infancy is a vulnerable period because individuals' immune systems do not mature completely until 2 years of age. **Pertussis** or **whooping cough**, a lower respiratory infection caused by *Bordetella pertussis*, develops in three stages: catarrhal stage with mucous membrane inflammation and symptoms resembling the common cold, paroxysmal stage with prolonged coughing sieges, and recovery stage of possibly several months. Gummy mucus forms and impedes ciliary action of the tracheal epithelium, potentially leading to ciliated epithelial cell death. Transmission by aerosol droplets from an infected infant occurs during the paroxysmal stage. Prevention by vaccination is begun at 2 to 3 months of age.

Topic Test 1: Bacterial Pathogens Transmitted Primarily by the Airborne Route

True/False

1. A set of disease symptoms in a host can be caused by only one bacterial species.

2. A particular bacterial pathogen can produce only one type of disease.

3. Humans in building their machinery can create new habitats for potentially pathogenic organisms.

Multiple Choice

4. What is the preferred way of controlling or preventing pertussis and diphtheria?
   a. Vaccination
   b. Penicillin
   c. Sanitizing the air with chlorine
   d. Cranberry juice
   e. Vitamin C supplements

5. Contraction of *L. pneumophila* differs from contraction of *S. pneumoniae* in what way?
   a. No fever occurs

b. Heart is involved in case of *L. pneumophila* and stomach in case of *S. pneumoniae*

c. Ticks spread *L. pneumophila*

d. Human to human transmission does not occur for *L. pneumophila*

e. Nosocomial infection

Short Answer

6. What environmental stresses do airborne pathogens face during transmission from one host to another?

Topic Test 1: Answers

1. **False.** *S. pneumoniae*, *K. penumoniae*, and *Haemophilus influenzae* induce similar forms of pneumonia in the host but are very different organisms.

2. **False.** *S. pyogenes* elicits numerous pathogenic responses by the host, depending on the portal of entry and the time during infection the pathologic event occurs.

3. **True.** Evaporative water chillers on large air-conditioning units provided a new habitat for *L. pneumophila*.

4. **a.** Prevention is always less stressful on the patient and more dependable than antibiotic therapy.

5. **d.** *L. pneumophila* spreads in humid air chilled in iron-containing equipment, not in droplets from coughing.

6. Airborne microorganisms, pathogenic or not, face desiccation, temperature fluctuations, and possible starvation if they are away from a host for very long.

TOPIC 2: BACTERIAL PATHOGENS TRANSMITTED PRIMARILY BY CONTAMINATED WATER OR FOOD

KEY POINTS

✓ *How can transmission of these pathogens be interrupted?*

✓ *What effect do the exotoxins produced by some of these pathogens have on humans?*

✓ *How do these microorganisms manage to remain in the intestinal tract of hosts?*

The **portal of entry** of the body for these microorganisms is the intestinal tract. They are introduced during eating or drinking of contaminated food or beverage (**food poisoning**). Some organisms produce exotoxins that damage intestinal epithelial cells and need not be present (**food intoxication**). Other organisms must multiply within the host's intestinal tract to have their effect be apparent (**foodborne infection**). To withstand the forces in the intestine, microorganisms have rods (**I pili**) or **adhesins**, mucopolysaccharides and glycoproteins to bind tightly to receptors on epithelial cells of the intestine. Many of these microorganisms spread by the four "F's": food, fingers, feces, and flies. If food or water quality is questionable, the following traveler's warning should be heeded: "boil it, peel it, cook it, or forget it."

Similar symptoms may be initiated by more than one species of bacteria, as shown in infections by *Vibrio cholerae*, *Shigella* species, and exotoxin-producing strains of *Escherichia coli*. Each of these organisms produces an exotoxin that directly or indirectly affects water retention by intestinal epithelial cells. The patient becomes dehydrated. *Shigella* and *Vibrio* are contracted by consuming contaminated seafood or drinking sewage-contaminated water appearing in epidemics accompanying displacement of large populations to refugee camps in wars or after natural disasters, such as earthquakes and large-scale flooding. *Escherichia* usually is spread through improperly prepared or insufficiently cooked meat. Fluoroquinolones are used in treatment of all three pathogens along with replacing the water and electrolytes lost in the diarrhea.

Nonhygienic preparation of food followed by inadequate storage of that food above 4°C may lead to **food poisoning** by *Staphylococcus* or *Salmonella* species. Staphylococcal food poisoning arises from multiplication of *Staphylococcus* on the food and release of exotoxins. **Intoxication** occurs when the food is eaten. Rapid onset of symptoms occurs 2 to 3 hours after eating the contaminated food, including nausea, vomiting, fever, and diarrhea. Recovery occurs in 72 hours. Fluids and electrolytes may have to be replaced if vomiting and diarrhea are severe. *Salmonella* food poisoning involves an **infection**, and symptoms do not appear for 12 to 48 hours, depending on the quantity of food eaten and its level of contamination. The organisms must multiply and invade the intestinal mucosa, producing an enterotoxin and a cytotoxin that destroy mucosal cells. Symptoms resemble those of staphylococcal food poisoning, and patients may require water and electrolytes to relieve the effects of dehydration in severe cases. Improperly treated drinking water and food prepared by infected cooks are the major sources of **typhoid fever** caused by *Salmonella typhi*. The microorganisms colonize the epithelium of the small intestine, penetrate it, and colonize lymphoid tissue, liver, and gallbladder. The patient develops high prolonged fever, headache, abdominal pain, anorexia, and malaise. Patients may continue to shed *S. typhi* for prolonged periods although they are symptomless **carriers**. The high fever is potentially lethal. Proper hygiene is the best method of managing typhoid fever. Carriers should receive antibiotic therapy to clear the pathogen from their system.

Improperly canned food may result in a deadly form of food poisoning, **botulism** caused by *Clostridium botulinum*. When the containers of cooked food are sealed hot, anaerobic conditions are created in the food in which endospores of *C. botulinum* can germinate. The resulting vegetative forms use the food as a medium, producing a neurotoxin, **botulinum toxin**. Cooking the food at high enough temperatures long enough kills the spores. Most cases of botulism occur from eating home-canned food. Rarely is commercially canned food involved in cases of botulism because commercial canneries are able to maintain cleanliness and proper cooking conditions more consistently than individuals involved in home canning. Because *C. botulinum* may produce hydrogen sulfide, a gas, in its metabolism, the presence of hydrogen sulfide, which has a strong odor, or bulging cans are warnings the contents may be contaminated.

Topic Test 2: Bacterial Pathogens Transmitted Primarily by Contaminated Water or Food

True/False

1. Consumption or use of purified chlorinated water in food preparation should eliminate food- or waterborne diseases.

2. All food- and waterborne diseases result from massive proliferation of certain microorganisms in the intestinal tract.

Multiple Choice

3. *Staphylococcus* and *Salmonella* food poisonings are associated often with summertime picnics because of
   a. unhygienic food preparation.
   b. eating food that is normally thoroughly cooked, raw, or partially cooked.
   c. failing to wash raw fruits and vegetables.
   d. allowing protein-rich foods to stay above 4°C for hours.
   e. not washing hands before eating.
   f. a and d above

4. Exotoxins of *Escherichia*, *Shigella*, and *Vibrio* cause extensive loss of what as a result of their disruption of ionic balances in intestinal epithelial cells?
   a. Nerve function
   b. Erythrocytes
   c. Water
   d. Vitamins
   e. Glucose reserves

5. Food poisonings associated with restaurants arise usually because
   a. their drinking water was contaminated with sewage water.
   b. they purchased contaminated raw fruits and vegetables.
   c. they had malfunctioning air conditioning.
   d. they failed to clean stovetops and countertops.
   e. employees failed to wash their hands thoroughly after contacting fecal matter.

Short Answer

6. What procedures can one do to minimize contracting one of the food- or waterborne diseases?

Topic Test 2: Answers

1. **False.** Contamination can occur also from unhygienic food handlers, contaminating the food with dust or soil, and improper refrigeration of stored food.

2. **False.** Ingestion of exotoxins produced by some microorganisms accounts for many food poisonings.

3. **f.** The food is contaminated by unclean preparation, but poor refrigeration is needed for the microorganisms to multiply sufficiently in the food.

4. **c.** Diarrhea results from water accumulation in the intestines.

5. **e.** An infected, but probably healthy, individual transfers his or her fecal organisms on his or her hands to the food being prepared.

6. First, make sure the water for drinking or washing food and eating utensils has been separated from sewage and chlorinated. Second, make sure food handlers have clean

hands. Third, make sure that soil and dirt are washed from food, and food preparation surfaces are clean. Last, stored food should be kept under conditions minimizing bacterial growth.

TOPIC 3: BACTERIAL PATHOGENS TRANSMITTED PRIMARILY BY DIRECT CONTACT

KEY POINTS

✓ *Why are these pathogens restricted to direct contact for transmission?*

✓ *How are infants infected by pathogens that are supposed to be transmitted sexually?*

Microorganisms requiring direct contact for transmission dehydrate easily, cannot destroy hydrogen peroxide (a byproduct of oxidative cellular respiration), are obligate intracellular parasites, or some combination of these limitations. Because sexual activity is accompanied by a prolonged period of contact and the sexual organs provide the limiting conditions required by these organisms, they are among the most widely occurring ones and induce sexually transmitted diseases or **venereal diseases**. Venereal diseases could be greatly reduced by the following procedures: public education about these diseases, prompt adequate treatment of new cases and all their contacts, sexual hygiene, monogamy, and use of condoms to prevent contact. **Nongonococcal urethritis** and **gonorrhea** are the most common bacterial infections of humans, ranking only behind the common cold in prevalence.

Chlamydia trachomatis is an obligate intracellular parasite needing direct contact for transmission. Infection in adults is primarily in the urethra or the vagina, leading to **nongonococcal urethritis**. *C. trachomatis* elicites an intense inflammatory response. In males, the infection is usually confined to the urethra. In females, with their shorter urethra the infection can spread to the bladder and the ureters or into the uterus and fallopian tubes, causing **pelvic inflammatory disease** and leading to miscarriage, stillbirth, eye infections, or infant pneumonia. Detection is by elimination of other possible agents, culturing the chlamydia in cell monolayers, or by observing serologically chlamydial antigens in urine.

Neisseria gonorrhoeae, the agent of **gonorrhea**, triggers a strong inflammatory response and is phagocytized by the phagocytic cells but may not be killed. *N. gonorrhoeae* may be detected by examining stained urethral discharges or vaginal scrapings for phagocytized gram-negative diplococci. In males, the infection begins in the urethra and may spread to the bladder, ureters, and kidneys or from the urethra to the vas deferens. Inflammation in the vas deferens may seal this tube and cause sterility. In females, the infection may spread through the urinary tract (rare) or into the reproductive tract. The urethra linings may be infected similarly to the male's. Usually, infection starts on the cervix when the organism is transmitted in the semen during coitus. The infection may spread into the uterus and fallopian tubes. Inflammation in the fallopian tubes may block them partially, enough to prevent passage of an egg but not of sperm. Fertilization may occur, but the egg will not reach the uterus and implants in the lining of the fallopian tube. This **ectopic** implantation will kill the mother unless an abortion is performed. Infants may contract **eye infections**, leading to blindness, in vaginal births.

Treponema pallidum, the agent causing syphilis, follows a very complicated progression that lasts over many years. After direct contact with a syphilic lesion, usually during coitus, an individual develops a small ulcer, known as a **chancre**, most frequently on the penis or scrotum or in the vaginal mucosa. This sore spontaneously heals in 1 to 2 weeks. Secondary skin lesions of various types that contain highly contagious spirochetes may occur over the body 6 weeks to 6 months after healing of the chancre and heal spontaneously in about 2 weeks. Repetitions may occur at 4- to 12-week intervals for 2 or more years. A latent stage that can be detected serologically by antibody to *T. pallidum* is entered. The spirochete has been detected in nerve cells during the latent period. Years later, the final stage appears with **gumma**, degenerative lesions of skin, bone, and nervous system, leading to death. *T. pallidum* may be transmitted transplacentally to fetuses. In fetuses, gross developmental defects are induced. Surviving fetuses are very abnormal. Penicillin has been very effective in treating syphilis at any stage but the terminal one.

Endospore-producing *Bacillus anthracis* is not an environmentally limited microorganism. Anthrax can be spread by contact with an infected person or by contact with materials that carry the spores, such as animal skins and wool. **Cutaneous anthrax** occurs when the organism gets into cuts or scrapes. Inhalation of endospores results in **pulmonary anthrax** (**Woolsorter's disease**), but endospores localized in the intestines produce **gastrointestinal anthrax**. The initial lesion is a large darkened ulcer, the **eschar**, accompanied by headache, fever, and nausea from exotoxins released. Penicillin and/or streptomycin are used to treat anthrax.

Topic Test 3: Bacterial Pathogens Transmitted Primarily by Direct Contact

True/False

1. Direct contact between infected host and another host must transfer many of these bacterial pathogens because they are intolerant of desiccation or aerobic conditions.

2. Fewer sexual partners and use of condoms would limit greatly the incidence of venereal diseases.

Multiple Choice

3. Detection of latent stage syphilis is by what method?
   a. Observation
   b. Serology
   c. Culture on an egg-blood agar medium enriched in iron
   d. Injecting mice and examining for chancres
   e. Collecting vaginal smears or penal pus secretions

4. Which of the following symptoms is *not* observed in gonorrheal infections?
   a. Infantile eye infections
   b. Urethral or vaginal pus secretions
   c. Ectopic pregnancy
   d. Bladder infections
   e. Cutaneous eschar

5. *C. trachomatis* is able to penetrate female bladders more frequently than male bladders because
   a. females have shorter urethras.
   b. males urinate harder and flush the pathogen out of the urethra.

c. males tolerate pain better.

d. female reproductive and urinary systems are separated.

e. None of the above

Short Answer

6. What steps can be pursued to reduce venereal diseases?

Topic Test 3: Answers

1. **True.** Only by direct contact can many of these pathogens pass from one host to another quickly enough to avoid lethal environmental conditions.

2. **True.** The first act would reduce exposure to an infected individual, whereas the second act interrupts the direct contact.

3. **b.** Only patient antibody to *T. pallidum* is readily detectable in this stage of syphilis.

4. **e.** This results from cutaneous anthrax.

5. **a.** Less distance must be traversed to reach the bladder.

6. Public education, treatment of patients and their contacts, sexual hygiene, monogamy, and use of condoms.

TOPIC 4: BACTERIAL PATHOGENS TRANSMITTED PRIMARILY BY ARTHROPODS AND THROUGH WOUNDS

KEY POINTS

✓ *How does one interrupt arthropod-borne infections?*

✓ *What are the sources of microorganisms entering through wounds?*

✓ *What methods or precautions can be followed to reduce wound-initiated infections?*

These diseases fall into two groups: One group where an arthropod **vector** carries the pathogenic microorganisms from host to host and the second group where the pathogen enters the host passively through breaks in the skin.

Arthropod vectors are limited to the blood-sucking insects and arachnids. In **plague** and **Lyme disease**, humans are opportunistic meals for the arthropods involved. **Rocky Mountain spotted fever** can be transmitted by **transovarian passage** through the eggs to the next generation of ticks. Infected ticks pass the disease to humans in their bite. The pathogens transmitted by insect vectors may harm their insect host. Plague kills both humans and fleas. The organisms of Lyme disease or Rocky Mountain spotted fever have little or no effect on their arthropod hosts.

 Mice and deer are the natural reservoirs for **Lyme disease**; humans are opportunistic hosts invading areas of hungry ticks. Lyme disease has three phases: a spreading ringlike rash accompanied by flulike symptoms, a disseminated stage occurring weeks or months later having neurologic and cardiac involvement and arthritis of major joints, and a terminal stage appearing years later and showing symptoms resembling Alzheimer's disease or multiple sclerosis. Prevention of tick bites by clearing and burning tick habitats near housing areas, wearing DEET (diethyltoluamide) containing repellants, and removing ticks from the body after being in high-risk areas controls Lyme disease best.

Rocky Mountain spotted fever is transmitted from wood or dog ticks infected with *Ricksettsia rickettsii* that triggers a characteristic rash and a high spiking fever. Serologic tests are used to diagnose the disease. When ticks feed, they inject an irritating anticoagulant and deposit feces contaminated with rickettsia near the wound. Humans introduce the rickettsia into the wound when they scratch the wound. Chloramphenicol and chlortetracycline are used in treatment. Wearing DEET repellents when moving in high-risk areas will reduce the chance of tick bites.

A **plague** epidemic begins with a die-off of rats and mice, the natural reservoirs of the causal agent *Yersinia pestis*. As the rodents perish, starving fleas feed on humans, the closest opportunistic meal for the flea. Fleas transfer *Y. pestis* that has been multiplying in the gut wall of the fleas to human blood. *Y. pestis* is quickly phagocytized and taken to a nearby lymph node, usually one in the groin. Outer membrane proteins on the *Yersinia*, encoded by a plasmid, block killing of *Y. pestis* in the phagocytic vesicles. *Yersinia* spreads to other phagocytes. The lymph node enlarges greatly to form a **bubo**, the origin of the name **bubonic plague**. Continued multiplication of the bacteria leads to overgrowth in the lymph node and release into the blood. Colonization of the lungs leads to shedding of *Y. pestis* in droplets. Fifty to 70% of individuals with lung infections die within 3 to 5 days. Because of the rapid progression of the disease, particularly in pneumonic plague, serologic techniques must be used to detect *Y. pestis*. Slower cultural methods can be used to confirm diagnosis. Prevention is achieved by destroying habitats for mice or rats near one's home and using insecticides to kill fleas.

Wounds, particularly large massive ones, offer a passive way for microorganisms to breach the body's outermost defenses. Blood supply to the area is disrupted, shackling other lines of defenses that rely on blood flow to reach the damaged tissue. Oxygenation of the tissue is reduced also, permitting an anaerobic environment to develop. *Clostridium tetani* and *Clostridium perfringens*, endospore-bearing widely dispersed soil organisms, can exploit wounds. *C. tetani* produces tetanospasmin, an extremely potent neurotoxin, whereas *C. perfringens* synthesizes a group of degradative enzymes and cytotoxins that digest the tissue in advance of the spreading bacterial colony. **Tetanus**, muscular spasms, induced by tetanospasmin leads to prolonged excruciatingly painful death. Vaccination with a chemically modified form of tetanospasm (tetanus toxoid) prevents intoxication. **Gas gangrene** caused by *C. perfringens* requires surgical debridement or removal of a limb above the infection followed by use of penicillin or sulfa drugs. Vaccination is not useful in this case because of the number of cytotoxins released and because the organism colonizes tissue with poor blood supply.

Topic Test 4: Bacterial Pathogens Transmitted Primarily by Arthropods and Through Wounds

True/False

1. Wounds provide a path for opportunistic microorganisms to bypass the outermost defenses of an individual.

2. One technique for interrupting the spread of diseases, such as plague or Lyme disease, is to reduce or eliminate the arthropod vectors.

Multiple Choice

3. Ticks that spread Rocky Mountain spotted fever do not require feeding on infected mammals to transmit the causal agent because
    a. they contract the ricksettsia from contaminated plant juices.
    b. transovarian transmission of the ricksettsia occurs.
    c. the ricksettsia is carried on hairs of fleas feeding on fecal matter.
    d. the ricksettsia is transferred by flea cannibalism.
    e. None of the above

4. For *C. tetani* and *C. perfringens* endospores to initiate colonization of wound tissue, what must be true of the wound tissue?
    a. Vascularized to ensure removal of carbon dioxide generated by microbial respiration
    b. Large enough to become anaerobic
    c. Enriched in iron required by the clostridia
    d. Rich in fatty acids to supply precursors for mycolic acid synthesis
    e. All of the above

5. Why are circulating antibodies against *Y. pestis* so ineffective in generating kill of the organism?
    a. The pathogen is protected by the blood-brain barrier.
    b. T_H lymphocyte activity is blocked.
    c. Phagocytized *Y. pestis* are not killed within phagocytic vesicles.
    d. Infected regions are surrounded by impervious cells with thick membranes.
    e. None of the above

Short Answer

6. If one must go into areas where ticks are abundant, what can one do to reduce the risk of contracting Lyme disease or Rocky Mountain spotted fever?

Topic Test 4: Answers

1. **True.** By colonizing these areas, the pathogens already have bypassed the tight outer body layer, and if blood supply is disrupted to the wound area, pathogens may escape the secondary response also.

2. **True.** Because arthropod vectors are necessary to transport these pathogens from one mammalian host to another, removing the vector would interrupt spread of the pathogens.

3. **b.** The ricksettsia penetrate eggs and thus can pass from one generation of ticks to another, perpetuating the parasite population without a mammalian alternate host.

4. **b.** These organisms are obligately anaerobic.

5. **c.** Plasmid-encoded outer membrane proteins apparently either prevent lysosomal fusion with the phagocytic vesicle or fusion does occur, but the lysosomal enzymes cannot digest the outer membrane proteins, leaving the pathogen still viable.

6. Wear an insect repellent containing DEET that is very repulsive to ticks and check body and clothing for ticks after one's excursion.

APPLICATION

Sometime around 1900, Mary Mallon emigrated from her native Ireland to New York. Her profession was that of a cook, and she was considered a good cook. She worked for many of the wealthy families in New York and on Long Island. She sometimes traveled with families to their New England vacation homes. At that time typhoid fever was endemic in the United States, but virtually all cases were found in lower economic neighborhoods where sewage mixed with drinking water. Thus, the occurrence of typhoid fever among wealthy families and their live-in servants was most unexpected. Careful investigation of commonalties in these cases showed that Mary Mallon had worked as a cook for them. The New York Health Department after a 6-month search finally caught up with Mary, now known as "typhoid Mary." Microbial examination of her feces showed she was excreting *Salmonella typhi*, the agent of typhoid fever. Despite this, Mary appeared to be a healthy, though overweight, individual. Some individuals who recover from typhoid fever continue to harbor the pathogen in their gallbladders. As with other microorganisms of the mouth, throat, and intestine, *S. typhi* and Mary had reached a compromise state where both could survive without harming each other. Mary was offered two choices: have a gallbladder removal or cease being a cook. The first was rejected because operations in the early 1900s were rather risky. She protested she could not do the second because that was her livelihood. She also suspected she was being persecuted. She was kept in confinement for 3 years, working as a hospital laundress. She was released, but changed her name to Brown and returned to cooking. Typhoid fever followed her again. Finally, she was incarcerated again, this time for life at Riverside Hospital in New York.

Chapter Test

True/False

1. Parasitic microorganisms are always pathogenic.

2. Some pathogens may be transmitted differently during different stages of infection.

3. Each pathogen initiates a unique set of symptoms in a host.

Multiple Choice

4. Which of the following procedures is *not* useful in preventing water- and foodborne infections?
   a. Thorough washing of hands
   b. Proper refrigeration of prepared food
   c. Spraying with DEET-containing repellent
   d. Separation of sewage and drinking water
   e. Chlorination of water

5. An apparently healthy individual who nonetheless transmits a disease is known as what?
   a. Vector
   b. Patient
   c. Parasite
   d. Carrier
   e. Operator

6. Cell surface substances on *M. tuberculosis* and *Y. pestis* aid these organisms in survival in a host by doing what?
   a. Providing nutrition when the bacteria are within macrophages
   b. Blocking the action of lysosomal enzymes in some manner
   c. Inhibiting activation of T_H lymphocytes
   d. Blocking processing of antigens by antigen-presenting cells
   e. Blocking the steps in the binding of complement proteins

7. *Staphylcoccus* food poisoning differs from *Salmonella* food poisoning because of what?
   a. One is found on protein-rich foods, the other on carbohydrate-rich foods.
   b. One has a hormonal effect, the other converts fats to organic acids.
   c. One is hydrophilic, the other hydrophobic.
   d. One results from an intoxication, the other from an infection.
   e. None of the above

8. *Vibrio cholerae*, *Shigella* species, and *E. coli* strains affect their host in what way?
   a. Exotoxins block maintenance of ionic balance, resulting in severe diarrhea.
   b. They trigger inflammatory reactions leading to destruction of intestinal mucosa.
   c. They colonize the gallbladder because of back flushing up into the bile duct.
   d. They trigger an autoimmune response.
   e. They are transmitted by transovarian route.

9. Many microorganisms associated with humans are opportunistic pathogens that become pathogenic when their hosts experience what?
   a. Dehydration
   b. Allergy
   c. Stress
   d. Relaxation
   e. Vaccination

Short Answer

10. What are the four "Fs" for transmitting food- and waterborne pathogens?

11. What makes treatment of *C. perfringens* infections so difficult?

Essay

12. Compare and contrast *T. pallidum* and *B. burgdorferi* infections.

Chapter Test Answers

1. **False**

2. **True**

3. **False**

4. **c**

5. **d**

6. **b**

7. **d**

8. **a**

9. **c**

10. Food, fingers, feces, and flies

11. The organism secretes a number of degradative enzymes and cytotoxins that destroy tissue, including vascular tissue, to provide oxygen for cellular repair and antibiotics to kill the pathogen. Surgical removal of dead or dying tissue and ample use of topical antibiotics will control *C. perfringens*.

12. Both organisms induce an acute phase, a latent phase of many years, and a terminal phase with gross physiologic dysfunction. *T. pallidum* has two types of acute phases. Both organisms enter the nervous system and create similar neurologic dysfunction. Both are sensitive to normal atmospheric oxygen levels.

Check Your Performance

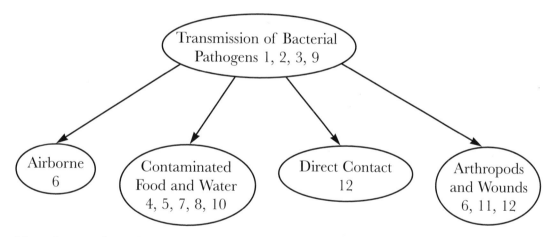

Note the number of questions in each grouping that you got wrong on the chapter test. Identify areas where you need further review and go back to relevant parts of this chapter.

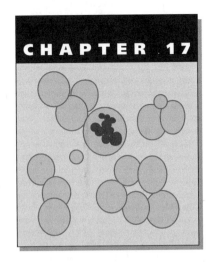

CHAPTER 17

Infectious Diseases Caused by Viruses

Viruses are obligate intracellular parasites, using their host's metabolism for their own needs. The viral genome can become incorporated into the host cell genome. To initiate infection, the virus must attach to a receptor on the host cell and stimulate the cell to bring the virus into its cytoplasm by pinocytosis. Because the viral pathogen takes over host cell metabolism, developing therapies to combat viral infections has proven difficult. Existing chemotherapeutic agents for viral infections slow viral multiplication, penetration, or maturation but do not clear the virus. The host's immune system must clear the virus. Compromising a host's immune system leaves the host very vulnerable to viral infections. Most treatments of viral infections are directed toward minimizing the physiologic discomfort experienced by the patient. Viral infections are prevented best by **vaccination**.

ESSENTIAL BACKGROUND

- Life cycles of bacteriophages: lytic and lysogenic (Chapter 9, Topic 1)
- Animal viruses (Chapter 9, Topic 3)
- Unusual viruses or viral-like agents (Chapter 9, Topic 4)

TOPIC 1: VIRUSES TRANSMITTED PRIMARILY BY THE AIRBORNE ROUTE

KEY POINTS

✓ *When should vaccination be used and what should be used for a vaccine?*

✓ *What determines which host and which cells viruses will invade?*

Because they are obligate intracellular parasites, viruses must move directly from one host to another. Their very small size makes dispersal in **droplet nuclei** efficient. Because viruses are not complete cells, dehydration and nutrient storage are not problems. Person to person spread through the air is by droplets produced when infected hosts cough, speak, or sneeze.

One of the most contagious diseases of childhood is **chickenpox (varicella)** characterized by an itchy red rash that fills with pus, ruptures, and forms scabs. Healing is spontaneous and reinfection does not occur. Spread is by droplets from the mild cough in the early stages of infection. The virus may reactivate in immunocompromised individuals, spread down the sensory nerves, and form painful vesicles on the skin called **shingles (zoster)**. No specific therapy is available, although drugs, such as acyclovir, lessen the intensity of the outbreak. Factors, such as age, AIDS, and physiologic or psychological stresses, seem to be most important in creating the immunocompromised state. Vaccines are available and recommended for infants. Quarantine has been used to try and isolate infected individuals, but it has not been effective. Individuals become contagious before they develop the characteristic rash.

Measles (rubeola) is contracted from droplet nuclei received in the respiratory tract or conjunctiva of the eye. The first symptoms resemble a common cold or the flu. Three to 5 days later, a red rash of small raised spots occurs on the face and trunk. Because CD 46 (membrane cofactor protein) found on many cell types is the attachment site for rubeola virus, the infection becomes systemic and may affect the eyes and nervous system. Characteristic bright-red **Koplik's spots** with bluish white specks in the center of the spot occur in the throat. There is no treatment beyond keeping the patient quiet and protecting the eyes if they become involved. Quarantines of infected individuals were used, but the most contagious phases had passed before the characteristic rash formed. Measles has become rare today in the United States because of mandatory vaccination of infants.

A number of different viruses targeting upper respiratory tract epithelial cells are spread by droplet nuclei. The symptoms of chills, fever, cough, congestion, headache, and malaise appear to be inflammatory responses. Congestion results from release of histamine from mast cells activated by IgE bound to viral epitopes. Probably the two most common viruses in this group are **influenza** virus and **common cold** viruses. Rhinoviruses can be transmitted by direct contact with handkerchiefs contaminated with nasal secretions. Hands contaminated with nasal secretions can leave rhinovirus on doorknobs, towels, bedding, or other items that are touched. There is no treatment for these viruses except to relieve symptoms with drugs like antihistamines. Attempts to develop rhinovirus vaccines are just starting because common colds are relatively mild and there are a large number of distinct rhinoviruses. Influenza may be accompanied by secondary bacterial pneumonia that may be fatal. The severity of influenza epidemics varies. Studying infections in the Far East in 1 year predicts with reasonable accuracy which strains will be prevalent the succeeding year. Polyvalent vaccines against the two or three strains expected to be epidemic the following year are distributed. Failures at guessing do occur.

Topic Test 1: Viruses Transmitted Primarily by the Airborne Route

True/False

1. The best method of managing viral diseases in populations is by using vaccines to prevent disease occurrence.

2. The symptoms observed in viral upper respiratory infections stem primarily from host inflammatory responses.

Multiple Choice

3. An individual presenting fever, a dry cough, raised red spots on the face and neck, and Koplick's sores in the throat probably has contracted what?
   a. Common cold
   b. Influenza

c. Measles

d. Chickenpox

e. Shingles

4. A virus residing in an individual without symptoms is said to be what?

a. Hiding

b. Recurrent

c. Sneaky

d. Festering

e. Latent

5. Why is vaccination not attempted for control of rhinovirus infections?

a. There are too many strains to make a successful vaccine.

b. Antibiotic therapy is cheaper and more successful.

c. They only rarely cause disease.

d. Vaccination for diphtheria cross protects against rhinoviruses.

e. The infections are severe and warrant prevention.

Short Answer

6. Why has quarantining sick individuals not been successful in stopping spread of measles and chickenpox?

Topic Test 1: Answers

1. **True.** There is no antibiotic therapy for eliminating viruses for individuals; they must rely on their immune systems.

2. **True.** In particular, histamine triggers many of the "cold" symptoms, which is the reason for consuming or nasal spraying antihistamines.

3. **c.** Koplick's spots plus the red rash are key symptoms of measles.

4. **e.** Latent viruses can be reactivated to produce a noticeable set of symptoms.

5. **a.** There are about 119 known strains of rhinoviruses, but most humans recover spontaneously in 7 to 10 days from these types of infections.

6. In both cases, infected individuals are shedding viruses in droplet nuclei during coughing before the development of recognizable symptoms for these diseases, namely the rashes.

TOPIC 2: VIRUSES TRANSMITTED PRIMARILY BY CONTAMINATED WATER OR FOOD

KEY POINTS

✓ *In what ways do individuals contract virus infections transmitted by these methods?*

✓ *How can one prevent contracting these diseases?*

Symptomology between viral-induced gastrointestinal infections and bacterial infections acquired via food and water frequently is blurred. Orally acquired viruses attach to receptors on the intestinal mucosal cells, triggering intestinal inflammatory responses. Host immune response to the presence of viral antigens in or on mucosal cells leads to most symptoms. Most of these infections are self-limiting, and the virus is cleared. The virus may manage to spread to other tissues, presenting more serious problems. Like intestinal bacterial infections, the mechanisms of contracting the pathogen follow the four F's rule: food, fingers, feces, and flies. Hygiene when preparing food and the availability of sewage-free drinking water are the primary means of reducing these types of infections.

Acute viral gastroenteritis is the most common of the gastrointestinal infections in this class and is caused by four categories of viruses: rotavirus, Norwalk and Norwalk-like viruses, other caliciviruses, and astroviruses. Infection from these viruses is more common in cooler months in contrast to bacterial infections being more frequent in summer months. The reason is unclear. Symptoms are usually relatively mild diarrhea, headache, fever, and vomiting. Treatment is to relieve the patient's physiologic discomfort. Rotaviruses are serious in infants because of malabsorption, impairment of sodium transport, and diarrhea. Rotavirus infections are leading killers of infants, particularly in the nonindustrialized areas of the world where drinking water is frequently contaminated. Treatment is by relieving the symptoms by supplying fluids and electrolytes.

Hepatitis A (infectious hepatitis) is contracted often from eating raw or improperly cooked shellfish. Viruses are contained in the digestive system of these filter feeders. Symptoms, if they occur at all, are mild forms of those typical of gastrointestinal infections. Occasionally, viremia occurs and viruses begin reproducing in liver cells leading to jaundice. Virus is shed in feces after being released in bile. An estimated 40 to 80% of the U.S. population has serum antibodies to hepatitis A virus.

Poliomyelitis, polio, or **infantile paralysis** is caused by poliovirus, a very stable virus that remains infectious for long periods in food or water. The virus enters mucosal cells of the intestine or throat and progresses to tonsils, lymph nodes of the neck, and terminal portion of the small intestine. If infection persists, the virus enters the central nervous system, penetrating and destroying the anterior horn cells of the spinal cord, resulting in motor and muscle paralysis. Destruction of respiratory function and the resulting asphyxiation are primary causes of death. Two vaccines are available to neutralize poliovirus.

Topic Test 2: Viruses Transmitted Primarily by Contaminated Water or Food

True/False

1. Vaccinations are the only means of combating water- or foodborne viral infections.

2. Shellfish harbor hepatitis A viruses because they filter them from sewage-contaminated water in which they live.

Multiple Choice

3. Which of the following mechanisms represents a source of intestinal viruses?
   a. Fingers
   b. Food

   c. Feces

   d. Flies

   e. All of the above

   f. None of the above

4. Those persons who develop chronic cases of hepatitis A shed virus through what path?

   a. Viruses are trapped in mucus in the lung and expelled in phlegm from the mouth.

   b. Viruses are produced with saliva and mixed either with food and leave with feces or expelled in droplet nuclei during talking or sneezing.

   c. Viruses formed in liver cells flow with bile to the gallbladder and then into feces.

   d. Viruses invade the kidney and, when mature, are excreted in urine.

   e. None of the above

Short Answer

5. Why have no vaccines been developed and used against viruses inciting acute viral gastroenteritis, but great effort was expended to develop the Salk and Sabin vaccines for polio?

Topic Test 2: Answers

1. **False.** Acute viral gastroenteritis infections are either left to run their course (they are self-limiting) or, if diarrhea is severe, are treated with fluids and electrolytes.

2. **True.** This occurs because of the type of feeders shellfish are.

3. **e.** This is the four F rule of gastrointestinal diseases.

4. **c.** If these viruses reach the liver, they can attach and penetrate well, preventing efficient elimination by the immune system.

5. Infections induced by the four types of viruses generating acute viral gastroenteritis are minor and self-limiting or, if significant, are treated by fluid and electrolyte replacement until immune clearance occurs. On the other hand, poliovirus that establishes chronic infection (1% of the cases) destroys key neural cells, resulting in paralysis and possibly death.

TOPIC 3: VIRUSES TRANSMITTED PRIMARILY BY DIRECT CONTACT

KEY POINTS

✓ *What is a self-limiting tumor virus?*

✓ *Is coital contact the only path for contracting these viruses?*

✓ *Can infection by all viruses in this group be prevented by vaccinations?*

Those viruses sensitive to dehydration must come in direct contact with cells they will penetrate. Many are transmitted during sexual activity. Some among this group do not attach to surface cells and must be placed so they can reach deeper tissues in the host. Attaining those tissues is achieved most often when virus-laden fluids are deposited in wounds reached by intimate contact between infected and healthy hosts.

Cold sores or **fever blisters** (**herpes labialis**) and **genital herpes** are diseases involving two closely related viruses: herpes simplex virus type 1 (HSV-1) and herpes simplex virus type 2 (HSV-2). HSV-1 usually infects cells at the mucocutaneous junction of the lips or conjunctiva of the eyes. The virus may become latent in trigeminal and autonomic ganglia. HSV-2 usually infects genital skin or mucosa. It may become latent in the sacral ganglia. Stress can trigger activation and outbreaks of viral blisters. HSV-1 may occur on the genitalia, and HSV-2 can occur in the oral region, depending on the type of sexual activity of infected individuals. Infants can be infected at birth when passing by infected vaginal mucosal cells. In infants, the virus invades the central nervous system and may cause extreme neurologic damage. Several topical chemicals interfere with viral multiplication by disrupting deoxyribonucleic acid metabolism. They do reduce the extent and length of the infection, but the immune system must do the clearing.

Human papilloma virus (HPV) frequently cotransmitted with HSV-2 during sexual intercourse infects epithelial or mucosal cells of external genitalia, vagina, cervix, or rectum, producing self-limited tumorous growths known as **genital warts**. This virus appears responsible for cervical or perianal cancer. Therapy involves removing the growths by electrosurgery, cryosurgery, laser fulguration, topical application of podophyllum, or injection of alpha-interferon.

Infectious mononucleosis (**mono** or **the kissing disease**) induced by Epstein-Barr virus (EBV) is transferred in oropharyngeal secretions during mouth-to-mouth contact (kissing) or sharing of drinking bottles or glasses. The virus is phagocytized by monocyte-macrophage type cells in the lymph nodes, multiplies, and infects B lymphocytes. B lymphocytes proliferate and become atypical with odd vacuoles in them. Symptoms of a bad cold appear with swollen lymph nodes, fever, and intense fatigue. Recovery occurs after a prolonged 6-week period. EBV may become latent in B lymphocytes and years later may initiate a fatal malignant B lymphocyte lymphoma. Patience is the only available therapy. The disease is most common in 15 to 25 year olds, probably because they often stress themselves and have multiple kissing partners.

Wounds are necessary for the following viruses to be deposited in deeper tissues of the individual because they cannot attach to surface cells. These viruses are considered here because there is direct contact between an uninfected individual and an infected one. Wounds are either made at the time of contact or were present and covered with contaminated fluid deposited in the wound region by the infected individual. **Rabies** virus causes rabies or hydrophobia. The virus initially penetrates skeletal muscle cells. As the concentration of virus increases, the virus spreads to the nervous system. Immune response is slowed when the virus is sequestered in neurons. Prior to neural invasion vaccination can be used to control infection. The virus spreads throughout the nervous system. From the peripheral nerves in the salivary glands, the virus is released into the saliva. The resulting encephalitis leads to difficulty in swallowing, and hydrophobia is observed. Humans contract the disease from mammal bites. It is thought that any mammal may contract rabies. Persons in high-risk professions, such as veterinarians or laboratory personnel, are vaccinated. Because animal bites are not common in the population at large and because the virus is susceptible to immune response until symptoms appear, vaccination just after being bitten allows destruction of the rabies virus.

Some viruses target cells of the immune system. These viruses must enter the blood to reach their host cells; thus, there must be breaks in the mucosa or epithelial layer of cells for the viruses to establish infection. **Human immunodeficiency virus** (**HIV**) and **human T-cell lymphotropic viruses 1** and **2** (**HTLV-1** and **HTLV-2**) may be transmitted sexually if a virus-infected individual has coitus with a partner who has open lesions from gonorrhea, syphilis, nongonococcal urethritis, or herpesvirus infections. Although transmission can occur from either gender, males are more likely to infect females, probably because contaminated semen is deposited in a vagina previously infected with other STDs and remains in contact with lesions longer than the vaginal fluids would remain in contact with penile lesions. Anal intercourse provides another portal for infection because the rectal linings are torn, sometimes exposing blood vessels to virus in contaminated body fluids. All three viruses can be transmitted by blood exchange at birth and are given in human milk, leading to infantile infections. HTLV-1 targets leukocytes, HTLV-2 seeks B lymphocytes, and HIV penetrates T_H lymphocytes having CD-4 on their surfaces. HTLV-1 and -2 can remain dormant for many years and then start rapid multiplication, leading to **adult T-cell leukemia** or **hairy-cell leukemia**, respectively. HIV likewise may have little effect on infected individuals for prolonged periods. Apparently, replication in tissue macrophages and/or activation of T_H lymphocytes to an infection triggers a burst of virus production that infects surrounding T_H lymphocytes. Hence, the frequency of infections in part determines the rate at which HIV infections reach their terminal state (**AIDS**). AIDS arises because HIV infections destroy T_H lymphocytes, major stimulators of immune response. There is no treatment for these diseases, although reverse transcriptase inhibitors and protease inhibitors, particularly when taken together, slow progression of the disease, delaying greatly the onset of AIDS. Attempts to construct vaccines against surface proteins on HIV, a technique used successfully earlier for other infectious viruses, have failed. HIV mutates at a rapid rate so that by the time one develops a vaccine, it is ineffective because a new strain has formed and spread. Antibodies produced against the vaccine do not recognize epitopes on the mutated viruses.

Topic Test 3: Viruses Transmitted Primarily by Direct Contact

True/False

1. There are chemical therapeutic agents available to clear viruses from infected individuals.

2. HIV, HTLV-1, and HTLV-2 are able to cross the placental barrier and infect fetuses.

Multiple Choice

3. Which of the following organisms transmits rabies by bites?
   a. Chickens
   b. Raccoons
   c. Bass
   d. Mosquitoes
   e. Ticks

4. Infectious mononucleosis is transmitted frequently by
   a. vaginal intercourse.
   b. anal intercourse.
   c. animal bites.

d. kissing.

e. holding hands.

5. Which of the following viruses is associated with tumor formation?

a. HTLV-1

b. EBV

c. HPV

d. HTLV-2

e. All of the above

f. None of the above

Short Answer

6. What has reduced the frequency of rabies dramatically in the United States?

Topic Test 3: Answers

1. **False.** The antiviral agents currently available can slow the rate of viral spread, but only a person's immune system can clear the virus from an infected individual.

2. **True.** The mechanism by which this occurs is unknown.

3. **b.** Mammals are the carriers of rabies.

4. **d.** EBV is released into saliva.

5. **e.** In each case the virus appears able to insert its nucleic acid into a host chromosome, much like a bacterial prophage.

6. The annual vaccination of most domestic dogs and cats for rabies has reduced the chance of human contact with a rabid animal to a very low level. Most human cases of rabies worldwide were contracted from domestic dog bites.

TOPIC 4: VIRUSES TRANSMITTED PRIMARILY BY ARTHROPODS AND THROUGH WOUNDS

KEY POINTS

✓ *Which arthropods are the major vectors of viral diseases?*

✓ *How might wounding pass HIV, HTLV-1, and HTLV-2?*

✓ *How is the spread of arthropod-borne viruses restricted?*

Humans are seldom the natural reservoir for viruses transmitted from one vertebrate host to another by blood-sucking arthropods. It is unknown how these viruses affect their arthropod vectors beyond establishing permanent infections. Three distinctive syndromes have been associated with various viruses in this group: fevers with or without a rash and/or arthritis, encephalitis (inflammation of the brain), and hemorrhagic fever (fever with bleeding due to destruction of blood capillary and vessel linings). Depending on the symptoms evoked, arthropod-borne viruses may invade mucosal cells, brain cells, or linings of synovial sacs between bones or cells of the vascular endothelium. The viruses in this group are not all related and have in common only their mode of transmission. An example of the fever/rash/arthritis class is **Colorado tick fever** triggered by a *Coltivirus*. Ground squirrels, rabbits, and deer are the principal reservoirs. Infected humans develop symptoms of a severe head cold accompanied by a rash, intense muscular pains ("break-bone fever"), and photophobia. Therapy is limited to relieving symptoms because the disease is self-limiting. There are no vaccines. DEET-containing insect repellents are used to prevent tick parasitism.

Most of the arthropod-borne viruses in the United States induce **encephalitises** that are named for the geographic location of discovery and the vertebrate harboring the virus at discovery. Mosquitoes transmit **California encephalitis** (La Crosse virus) from chipmunks and squirrels, the natural vertebrate hosts. This form of encephalitis is rarely fatal. **Eastern equine encephalitis**, **St. Louis encephalitis**, **western equine encephalitis**, and **Venezuelan equine encephalitis** are more severe, with fatalities ranging from 3% for western equine encephalitis to 70% for eastern equine encephalitis. The first three of these diseases have birds as their vertebrate reservoirs. Chickens are placed in high-risk sites to determine if encephalitis-carrying mosquitoes are in the area. Only Venezuelan equine encephalitis has horses and rodents as its reservoirs. Horses are very susceptible to any of these severe encephalitises. Treatment is to relieve the high fevers. Neural destruction in the brain probably results from the intense immune responses to the presence of the virus and to the rapid spread of the virus once it is in brain cells. In addition to death, more common complications are epilepsy, paralysis, deafness, blindness, or mental retardation. Prevention of encephalitis is a four-pronged program: vaccinate horses and humans at high risk, spray mosquito-breeding grounds with malathion, use DEET-containing insect repellents liberally if going outside at dawn or dusk when mosquitoes are most active, and wear clothing to cover most of the body.

Hemorrhagic fevers rarely occur today in the United States, although in the early 1900s **yellow fever** occurred in southern United States. Control of mosquito populations by insecticides and draining stagnant ponds interrupted transmission by mosquito vectors. Monkeys are the natural reservoir. Yellow fever virus invades the gastrointestinal vascular system triggering massive hemorrhages, leading to "black vomit." The liver is invaded, producing jaundice from the release of liver bile into the blood. Most cases are milder and similar to a severe head cold. For travelers to the tropics where yellow fever is endemic, there is a vaccine that can be administered before traveling. Citizens of the United States experienced **dengue fever** virus transmitted by mosquitoes on the tropical South Pacific islands during World War II. Humans are the most prominent vertebrate host, although monkeys may serve as hosts in tropical forested areas. Symptoms are usually a severe cold with a high fever. If endothelial cells are invaded, massive hemorrhaging through body orifices can occur with loss of blood pressure that, if drastic enough, leads to circulatory collapse. Dengue is developing into a major problem throughout tropical regions of the world because of rapid urbanization without the proper infrastructure to prevent sewage and rainwater ponding, providing mosquitoes breeding grounds. Control of dengue fever requires drainage of mosquito-breeding ponds, use of insecticides to reduce mosquito populations, and vaccinations against the virus.

Wounds are important portals of entry for viruses that attach only to cells of deeper body tissues. HIV, HTLV-1, and HLTV-2 require blood-to-blood transfer. Direct contact between fluids of infected individuals with rectal tears or genital wounds may occur (see Topic 3). Materials contaminated by blood from infected individuals can transmit viruses. Hemophiliacs of both genders and of any age were presenting AIDS symptoms in the early 1980s at a time when only homosexuals were showing AIDS. This was confusing until it was demonstrated the hemophiliacs had received antihemophiliac factor from HIV-contaminated blood. Intravenous drug users of both genders develop AIDS from sharing needles contaminated with blood carrying HIV. Accidental needle punctures by health care providers serve as another form of wound-acquired HIV blood-to-blood transmission. HTLV-1 and HTLV-2 can be transmitted in the same ways as HIV.

Topic Test 4: Viruses Transmitted Primarily by Arthropods and Through Wounds

True/False

1. Mosquitoes are the most important vectors of arthropod-borne viruses.

2. Removing the breeding grounds of mosquitoes has dramatically reduced the occurrence of diseases, such as equine encephalitis and yellow fever.

Multiple Choice

3. Which of the following organisms may serve as reservoirs for arthropod-borne viruses?
   a. Ground squirrels
   b. Chipmunks
   c. Birds
   d. Humans
   e. Monkeys
   f. All of the above
   g. None of the above

4. In addition to the means of contracting HIV or HTLV infections by sexual activity, how else might one contract HIV or HTLV?
   a. Food contaminated with fecal matter
   b. Mosquitoes
   c. Pricks from contaminated needles
   d. Dirty eating utensils and glasses
   e. Toilet seats

5. The "black vomit" and bleeding seen in yellow fever and dengue hemorrhagic fever occur because
   a. humoral coagulation proteins are incorporated into viral envelopes.
   b. endothelial cells are damaged.
   c. liver bile released into the blood digests membranes of endothelial cells.
   d. accumulation of water swells capillaries to bursting, leading to bleeding.
   e. None of the above

Short Answer

6. What can one do to reduce the risk of arthropod-borne viral encephalitis infections?

Topic Test 4: Answers

1. **True.** Only California encephalitis virus has a non-mosquito vector, a tick.

2. **True.** This phenomenon has been and is being repeatedly demonstrated in southeastern United States.

3. **f.** Although the viruses vary, each of these animals serves as a reservoir for an arthropod-transmitted viral infection.

4. **c.** This is the primary means of transmission among health care providers, homophiliacs, and illicit intravenous drug users.

5. **b.** Endothelial cells infected with flaviviruses are damaged probably by the intense inflammatory response triggered, and the endothelial linings become porous as a result.

6. One can follow a four-pronged program: vaccinate horses and/or humans at high risk, spray mosquito breeding grounds with malathion, liberally use DEET-containing insect repellents when going into areas with large mosquito populations, and wear clothing covering most of the body to prevent bites if mosquitoes land.

APPLICATION

Early May 1993 was not a relaxed time on the Navaho Reservation in northwestern New Mexico. Four young adult athletic Navahos had developed seemingly typical influenza symptoms of fever, headache, and respiratory congestion. Three of the four died suddenly gasping for their breaths. What was alarming about these cases was the age and prior health of these patients. The Navaho have a history of severe respiratory infections besides influenza, including bubonic plague, *Haemophilus influenzae*, and viral pneumonia. Autopsy of the victims revealed lungs so filled with fluid they weighed twice their normal weights. This is not characteristic of any of the above respiratory illnesses. Attempts to isolate bacteria from lung tissue failed. Also, immunologic analysis of serum and lung samples from patients failed to show the presence of any of the viruses typical to the Navaho. By the first week of June, the number of similar cases had increased to 21 with 12 fatalities. By now the search to understand this disease included the Centers for Disease Control and Prevention, the New Mexico Department of Health, and physicians in the health clinics of the Navaho Nation. At first, phosphene, a rodenticide used against plague-carrying prairie dogs, was suspected. Two factors suggested that phosphene was not the lethal agent: very few toxic chemicals elicit fever and examination of the manufactured homes in which the early patients resided contained no evidence of phosphene. What was found at these houses was mouse feces.

In the meantime, blood samples from victims and patients were shipped to the Centers for Disease Control and Prevention in Atlanta. Using a large battery of antiviral antibodies, investigators narrowed their search to hantaviruses because of cross-reactions in patients' blood samples to this group of viruses. To confirm this discovery, patient blood was injected into mice who produced a strong antihanta virus antibody reaction. Using polymerase chain reaction (PCR) and other molecular biology techniques with patient blood samples, investigators found that indeed a new hantavirus was present. Hantaviruses were discovered first in Korea during the 1950s when 121 American soldiers died from a mysterious influenza-like disease. After 20 years of investigation, these first known hantaviruses were observed and shown to have field mice as their reservoir. Korean Hantaan virus had been observed in Baltimore, carried there from Korea apparently on freighters in international commerce. Field mice do not live on the Navaho Nation, however.

Conversations with Navaho shamans brought out the fact that a heavy winter snowpack had provided the moisture for the pinyon pine to produce a bumper crop of pinyon nuts and an explosion in the population of deer mice that feed on the nuts. At the end of the summer, the deaths from hantavirus spread over six southwestern states with 42 confirmed cases and a 62% mortality. Collection of mouse feces from dwellings of patients and capture of deer mice in the vicinity of those dwellings yielded from PCR and DNA analyses the presence of the same hantavirus found in the patients' specimens. Hence, this hantavirus appears to be airborne from dried deer mouse feces. The virus has been named after its site of discovery, the Muerto Canyon virus. (Muerto Canyon is Spanish for the Canyon of Death, an apparently very appropriate designation in this case.)

Chapter Test

True/False

1. All viruses are obligate intracellular parasites.

2. Viruses can penetrate any host cell, which is why they are systemic pathogens.

3. Because viruses are intracellular parasites, they can only be transmitted by direct contact.

4. Although drugs, such as reverse transcriptase inhibitors and protease inhibitors, slow virus penetration and multiplication, no known chemical therapy clears viruses from a host.

5. Some viruses require breaks in epithelial and mucosal tissue because they attach only to cells of deeper tissues, not to surface cells.

Multiple Choice

6. Many viruses can be controlled by
   a. vaccination.
   b. antibiotic therapy.
   c. nutrition.
   d. vitamin C pills.
   e. sunlight.

7. Many of the symptoms of viral infections, such as those seen with the common cold, dengue fever, and encephalitis viruses, occur as a result of
   a. penicillin hypersensitivity.
   b. complement fixation.
   c. inflammation.

d. secondary bacterial infections.

e. macrophage dysfunction.

8. How do viruses that attach to deeper tissues reach those tissues when they cannot attach or multiply in surface cells?

a. Enter the blood when punctures occur

b. Enter through ulcers on epithelial and mucosal surfaces

c. Placed in blood by blood-sucking arthropods

d. Rectal tears occurring during certain forms of coitus

e. All of the above

f. None of the above

9. Viruses of common cold, measles, or chickenpox spread through a population by what manner?

a. Dog bites

b. Needle sticks

c. Drinking water

d. Mosquito bites

e. Droplet nuclei from coughing

10. HIV is such a devastating disease because

a. extreme diarrhea and ion loss occurs.

b. inflammatory response destroys the central nervous system.

c. erythrocytes are produced uncontrollably.

d. T_H lymphocytes are destroyed.

e. motor neurons are destroyed.

11. Discussions in Topic 3 indicate the uninfected individual becomes infected when coming into contact with bodily fluids. Which of the following fluids are known to be free of viral contamination?

a. Saliva

b. Milk

c. Vaginal fluid

d. Semen

e. Fluid from blisters

f. All of the above

g. None of the above

Short Answer

12. One can reduce the risk of contracting hemorrhagic fevers by doing what?

13. Why is it possible to vaccinate against rabies virus successfully after being exposed to this virus?

14. What means lead to the transmission of gastrointestinal viruses?

Essay

15. How can one reduce the risk of contracting or passing HIV with a sexual partner?

Chapter Test Answers

1. **True**
2. **False**
3. **False**
4. **True**
5. **True**
6. **a**
7. **c**
8. **e**
9. **e**
10. **d**
11. **g**
12. Destroying potential breeding sites, wearing DEET-containing insect repellents, and getting vaccinated.
13. Rabies virus slowly multiplies in skeletal muscle cells and requires 2 weeks to several months before high enough virus concentration is reached to enter the nervous system. Thus, in the first days after exposure, the virus is vulnerable to an induced immune response that is triggered strongly by vaccination.
14. Fingers, food, feces, and flies.
15. Prevent spread by direct contact to preexisting lesions by using a condom. Reduce one's partners by practicing abstinence or at the most monogamy.

Check Your Performance

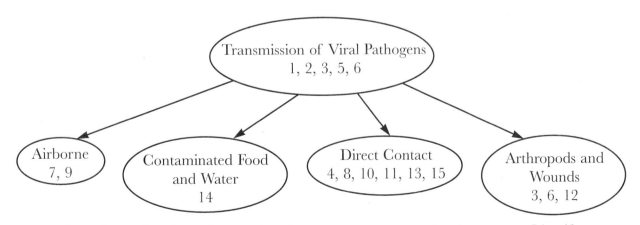

Note the number of questions in each grouping that you got wrong on the chapter test. Identify areas where you need further review and go back to relevant parts of this chapter.

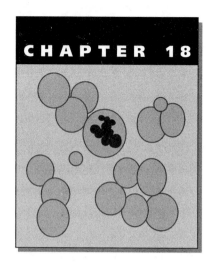

CHAPTER 18

Diseases Caused by Protozoa, Fungi, and Algae

These organisms are eukaryotic and primarily opportunistic parasites or produce toxins that are ingested. Mosquitoes transmit some widely, such as malaria. Very effective immune responses of humans and other mammals limit the ability of eukaryotic microorganisms to colonize potential hosts.

ESSENTIAL BACKGROUND

- **What are prokaryotes and viruses? (Chapter 1, Topic 2)**

TOPIC 1: OVERVIEW OF THE BIOLOGY OF FUNGI, PROTOZOA, AND ALGAE

KEY POINTS

✓ *What niches do fungi occupy in the biosphere?*

✓ *What characteristics are used to classify and identify fungi?*

✓ *What are the modes of protozoal reproduction?*

✓ *How does the protozoal life cycle affect its mode of transmission?*

✓ *What is the primary mode of action by which algae cause disease?*

Fungus is the Latin word for mushroom, a prominent reproductive structure on some forms of fungi. Microscopically, the eukaryotic fungi appear as elongated branched threads (**hyphae**) that form tangled masses (**mycelia**) or are spheroidal cells (**yeasts**) that sometimes are grouped in grapelike clusters when progeny cells remain attached to parent cells. Fungi release exoenzymes to digest macromolecules and then absorb the resulting degraded material ("slop"). None of them have chlorophyll or are autotrophic. Various fungi reproduce asexually, sexually, or both. Fungi are found primarily in soil wherever there is organic matter to be degraded or where they can form close mutualistic symbiosis with other organisms. Some are parasitic and can be pathogenic, particularly on plants. Approximately 90,000 fungal species have been described. Many more probably will be discovered, possibly as many as 1.5 million, especially as soil is examined more thoroughly.

Fungi play three major roles in the biosphere: conversion of dead organic matter into usable substances, mutual symbioses with many macroscopic and microscopic organisms, and parasitism

and pathogenesis, mainly on plants. A few fungi can colonize animals and become opportunistic
pathogens. As long as T lymphocyte functions are operating in birds and mammals, most fungi
seem to be restricted to bodily surfaces. Fungi establishing themselves in apparently healthy
hosts use the lungs as the main portal of entry. Fungal colonization is restricted by their
inability to successfully compete with bacteria of the normal flora on skin, mouth, and vaginal
linings.

Fermentation by fungi was the primary method most humans had for preserving food before the
introduction of refrigeration and still is used in many places in the world where refrigeration is
rare. Fungal metabolism is harnessed for commercial production of organic acids, such as citric
and gallic acids, and pharmaceuticals, such as cortisone, penicillin, griseofulvin, and the
immunosuppressive agent cyclosporine.

The vegetative structure of a fungus, the **thallus**, varies greatly in size, shape, and complexity.
The simplest thalli are the unicellular yeast or the multicellular linear or slightly branched
hyphae in which no division of labor occurs among the cells. **Hyphae**, elongated filaments, may
be **coenocytic** with no cell walls partitioning the hyphae or **septate** with regular partitioning
of the hyphal length. Septate hyphae have cross walls that are interrupted by a single large pore
or are multiperforated with cytoplasm connecting adjacent cells. Cell walls bounding hyphae
or yeast are very thick, containing chitin. Fungi share chitin as a supportive material with
arthropods.

Fungi are mostly facultatively aerobic chemoheterotrophs that can obtain some energy by fer-
menting glucose to ethanol. Fungi require carbohydrate or some reduced carbon source for
carbon, electrons, and energy, usually an inorganic nitrogen source, a balanced salt solution, and
water. They synthesize the amino acids, phospholipids, proteins, and vitamins they need for survival. With the exception of the yeast, biochemical techniques have not been a fruitful way of
recognizing fungal species. Obligately anaerobic fungi are found in the rumen of cattle and
other ungulates. Certain wood-rotting fungi are among the few organisms capable of hydrolyzing
cellulose and lignin found in woody plants. Some potentially pathogenic fungi on animals are
dimorphic, that is, they can change from a yeast (Y) form in the animal to a hyphal form (M)
external to the animal. This **YM shift** is triggered by changes in nutrition, temperature, CO_2
tension, and oxidation-reduction potential. Some plant-associated fungi show a MY shift, being
hyphal (M) on plant hosts and yeastlike (Y) in the environment external to the plants.

Reproduction in fungi may be asexual or sexual. The shape and distribution of reproductive
structures is the basis for identifying and classifying fungi except for yeasts. Many fungi have
demanding requirements before forming reproductive structures, particularly sexual forms.
Fungi demonstrating sexual reproduction also have asexual reproduction. One group of fungi
has asexual reproduction but no known sexual forms. Another group exhibits only vegetative
growth (hyphae) and no sexual or asexual structures.

Protozoa (Greek for "first animals") constitute a group of single-celled eukaryotic organisms
known as **protists** because they lack true differentiated tissues. Protozoa have no cell wall, no
photosynthesis, motility usually, and no fruiting bodies. Protozoa are classified into four groups
by their motility. The protozoal cytoplasm usually has two parts: the outer layer, **ectoplasm**,
for motility, protection, and feeding, and **endoplasm**, the inner portion, containing the nucleus
and mitochondria. All protozoa are heterotrophs. During their life cycle, all protozoa have a
trophozoite (feeding stage), whereas some form **cysts** when growth conditions become
unfavorable.

The adult stage of **sporozoans** (Apicomplexa) is nonmotile and obligately parasitic. Nutrients are absorbed in soluble form rather than being ingested by pinocytosis, phagocytosis, or swallowing by way of a gullet. After sexual reproduction, **sporozoites** contained in oocysts are transmitted between hosts. Asexual reproduction in host cells occurs by multiple fissions of sporozoites in the host cells (**schizogony**).

Ciliates (Ciliophora) possess cilia for motility and are the only protozoa that have two types of nuclei: diploid **micronucleus** for reproduction and inheritance and the polyploid **macronucleus** for production of mRNA during vegetative growth. Nutrients are ingested by way of a gullet.

Flagellates (Mastigophora) use flagella for motility. Some pathogens in this group form cysts that are transmitted. Otherwise, they are spread by insect vector or direct contact. Free-living pathogenic **ameba** (Sarcodina) form cysts that are transmitted. Infection targets the central nervous system (CNS) or the gastrointestinal tract.

Algae are photosynthetic eukaryotes characterized by diversity in size and color. The Euglenophyta (**euglenids**) containing the genus *Euglena* grow as heterotrophs in the absence of light. Algae are motile and have cells walls, except *Euglena*, which has a thick flexible membrane, a **pellicle**. Algae populating the large floating mass of oceanic microorganisms (**plankton**) produce most of the world's oxygen.

Topic Test 1: Overview of the Biology of Fungi, Protozoa, and Algae

True/False

1. Fungi are found in virtually every habitat explored but primarily decay organic matter in soil.

2. Protozoa are heterotrophs.

Multiple Choice

3. What is the feeding form of protozoa?
   a. Cyst
   b. Sporozoite
   c. Trophozoite
   d. Merozoite

4. Which of the following situations demonstrates mutualistic symbiosis between a fungus and another organism?
   a. Shelf fungus on the trunk of a fallen tree
   b. Lichens
   c. Yeast living in the human mouth
   d. Fungus isolated from a chlorotic leaf spot
   e. None of the above

Short Answer

5. What is the YM shift observed with some fungi?

6. What property is used to classify protozoa into groups?

Topic Test 1: Answers

1. **True.** All are chemoheterotrophic; hence, without organic matter they are unable to obtain the carbon, electrons, and energy needed to survive.

2. **True.** No protozoa have the ability to obtain their carbon from carbon dioxide or any other inorganic compound.

3. **c.** Trophozoite refers to the form in the protozoal life cycle during which nutrients are incorporated.

4. **b.** Fungus provides protection from environmental stresses, such as desiccation, whereas the associated algae or blue-green bacteria provide photosynthesis to make sugars for nutrition.

5. YM shift occurs where opportunistic fungal parasites grow in the yeast form on warm-blooded (homeothermic) organisms but as hyphae (mycelia) in the external environment.

6. The mode of motility is used to classify protozoa into groups.

TOPIC 2: INFECTIOUS DISEASES CAUSED BY FUNGI

KEY POINTS

✓ *What sites on mammals and birds can fungi colonize?*

✓ *What limits most fungi from colonizing mammals or birds?*

Mycoses are the group of fungal-induced infections of mammals and birds limited usually to surface tissues. Fungi can colonize immunocompromised mammals or birds well. Two factors seem to restrict fungal development: T lymphocyte populations of the potential hosts destroy invading fungi and these hosts have extensive populations of bacteria on their surfaces that compete very successfully with fungi. Attachment to the host is difficult for fungi, and the competition for nutrients is even more difficult. Few good antibiotics for treating fungal infections exist. Many compounds toxic to fungi are known but are also toxic to the hosts. Antibiotics too toxic for consumption may be useful topically on body surfaces.

Superficial mycoses limited to body secretions or hair occur mainly in the moist tropics. The fungi secretions appear to obtain the carbon and nitrogen sources they require from hair proteins and skin. Humidity of the air in which the infected individual lives may be a factor also. The hard stonelike growths on hair shafts are known by the Spanish word for stone, **piedra**: black piedra (*Piedraia hortae*) on hairs of the scalp, white piedra (*Trichosporon beigelii*) on the beard and mustache, and **tinea versicolor** (*Malassezia furfur*) brownish-red scales on the trunk, neck, face, and arms. Treatment is removal of scales with a cleansing agent or clipping the infected hairs. Good personal hygiene is the best prevention.

Dermatophytoses or **cutaneous mycoses** occur in skin cells, creating ringlike patches (ringworm). These common diseases are named after their site of occurrence: tinea cruris or jock itch on the groin and buttocks, tinea pedis or athlete's foot on the feet, and tinea unguium (onychomycosis) on the nails. Primary treatment is with topical ointments or fluids containing miconazole, tolnaftate, clotrimazole, or undecylenic acid.

Subcutaneous mycoses found among barefooted agricultural workers, particularly in the humid tropics, are caused by opportunistic fungi taking advantage of wounds on the feet and lower legs of individuals. The fungi do not appear to be well adapted to human or animal tissue because they are slow growing on the subcutaneous layers of the skin. Undoubtedly, nodulation and/or ulceration observed result from immune response to the presence of fungi in subcutaneous cells. In the tropics, **chromoblastomycosis** (*Phialophora verrucosa* or *Fonsecaea pedrosoi*) forms dark brown-pigmented nodules and **eumycotic mycetoma** or **fungal tumor** (*Madurella mycetomatis*) with destruction of subcutaneous tissue. **Sporothrichosis** (*Sporothrix schenckii*), common in the United States, is found on many horticultural plants, plant products, and in the soil. This disease is seen most in florists, gardeners, and forestry workers who work with prickly plants, such as roses, or plant debris, such as sphagnum moss and pine-bark mulch. Red papules and ulcers occur and may spread over the body. The three forms of infection are treated orally with 5-fluorocytosine or amphotericin B or by surgical removal.

The agents of **systemic mycoses**, primarily saprophytic organisms in soil and fecal matter, are able to colonize deeper tissues either through wounds or by spore germination in the lungs. They produce spores that are easily disseminated in winds, accounting for the lungs as major portals of entry. *Coccidioides immitis*, found widespread in dry alkaline soils of the Western Hemisphere, produces readily airborne arthroconidia that germinate in the lungs. The fungus experiences a YM shift and multiplies as a yeast with thick walls and endospores resulting in **coccidioidomycosis** (valley fever, San Joaquin fever, or desert rheumatism), reflecting the geographic distribution of cases and flulike symptoms. Most infections resolve themselves within a few weeks unless the patient has impaired T lymphocyte functions. Prevention is achieved by wearing a mask when working outdoors in areas where the fungus is endemic. *Cryptococcus neoformans*, a yeast, contracted from inhaling airborne yeast from dried pigeon droppings, is usually a mild pulmonary infection occurring worldwide and is usually self-limiting in persons with functional T lymphocytes. In immunocompromised patients, the yeasts may spread to skin, bones, viscera, and CNS. Detection of yeast in pus, sputum, or exudate smears in India ink along with culturing confirms diagnosis. No control or prevention measures exist, although cleaning up pigeon manure reduces airborne yeast. *Histoplasma capsulatum*, a mycelial saprophyte common in bird and bat manure, produces small microconidia easily dispersed in air currents over bird droppings or bat guano. Lungs are portals of entry with a mild flulike disease (**histoplamosis**). The yeast phase is cleared except in immunocompromised individuals where the disease may persist in macrophages in calcified nodules, not unlike tuberculosis. Masks worn when working in areas with bird droppings or bat guano and disinfection of dirty clothing reduce the risk of infection. Bats and humans appear to be the main hosts for *H. capsulatum*. Birds are carriers but have too high a body temperature to support *H. capsulatum* growth.

The last group of fungal pathogens is completely **opportunistic**, establishing invasive infections in immunocompromised individuals. Both *Aspergillus* species and *Candida albicans* are ubiquitous saprophytes, although *C. albicans* is frequently part of the normal microflora of the mouth, vagina, upper respiratory tract, and gastrointestinal tract of humans. *Aspergillus* species rarely invade beyond the surfaces of the respiratory tract and may form "fungus balls" that impede the air passages. Most damage results from the intense inflammation triggered by type I and type III hypersensitivities to *Aspergillus*. Diagnosis is by examination of specimens and culturing of the agent. *C. albicans* (**candidiasis**) invades almost anywhere in or on the body, depending on the immunocompetence of the individual, portal of entry, ease of dissemination, health of normal bacterial flora, and the hormonal state of the individual. Disseminated candidiasis has become a signal of immune dysfunction, particularly T lymphocyte dysfunction observed in HIV infections.

Topic Test 2: Infectious Diseases Caused by Fungi

True/False

1. Many fungal pathogens may become invasive if host T lymphocyte function is compromised.

2. Those fungi with YM shifts grow as yeasts on mammalian or avian hosts.

Multiple Choice

3. Fungi inciting subcutaneous mycoses invade their hosts by what portal of entry?
   a. Genitourinary tract
   b. Lungs
   c. Gastrointestinal tract
   d. Wounds
   e. Hair follicles

4. What is the major method for controlling fungal infections?
   a. Vaccines
   b. Replacing lost fluids
   c. Surgical removal
   d. Ultraviolet light
   e. Antibiotic therapy

5. Why apparently are superficial and cutaneous mycoses more common in tropical regions of the world?
   a. More solar damage to the skin
   b. Humidity
   c. Lack of cold weather
   d. Fewer bacterial competitors
   e. All of the above

Short Answer

6. What is the major portal of entry for fungi causing systemic infections? Why?

Topic Test 2: Answers

1. **True.** T_C and probably T_{NK} lymphocytes appear to attack and release cytotoxins that are particularly damaging to fungal cells.

2. **True.** Yeast phase is favored by high body temperature and increased CO_2 tension in the vicinity of infections.

3. **d.** These are opportunistic pathogens with little capability of penetrating the surface defenses of mammals.

4. **e.** No vaccines are available, and surgical scraping is used for one type of subcutaneous mycosis. Ultraviolet radiation affects skin more than the fungi. Diarrhea and vomiting are not symptoms of any mycoses.

5. **b.** Fungal spores germinate in warm humid conditions. In addition, sweating in those warm conditions adds moisture on the skin.

6. These fungi produce very small spores on decaying organic materials in soil. The spores readily become airborne and are breathed into the lungs where they find a warm moist habitat.

TOPIC 3: DISEASES CAUSED BY PROTOZOA AND ALGAE

KEY POINTS

✓ *What determines the mode of transmission of protozoal diseases?*

✓ *What are the most effective methods for control of protozoal infections?*

✓ *What are the major diseases caused by protozoa?*

✓ *What is the major mode by which algae cause human diseases?*

Diversity in the life cycles of pathogenic protozoa is responsible for the variety of modes of transmission of protozoal diseases. Knowledge of the reservoirs, sources, life cycles, and transmission modes is useful in developing effective control methods for protozoal diseases. Their control by antibiotics and chemotherapeutic agents is limited. Because they can undergo adaptive mutations and because some also have a sexual reproductive mode that increases genetic diversity, it has been difficult to develop effective vaccines.

Ameboid pathogenic protozoa form cysts during their life cycle, which may be transmitted to uninfected hosts. Amebic dysentery (**amebiasis**) caused by *Entamoeba histolytica* is widely distributed. Unsanitary conditions and the contamination of food and water supplies are sources of amebiasis. Ingested cysts lodge in the small intestine where trophozoites form. The trophozoites move to the large intestine and multiply. Some strains release exotoxin that dissolves tissue and penetrates into deeper intestinal mucosal layers, giving rise to ulceration of the intestinal lining, with bloody mucus-filled stool (**dysentery**). Control of amebiasis requires prevention of the contamination of food and water supply by cysts from infected individuals.

Free-living pathogenic protozoa *Acanthamoeba* species and *Naegleria fowleri* are naturally present in standing fresh or brackish water, wet soils and mud, ponds, hot springs, swimming pools, and hot tubs. Human infection by these amebae is rare. They occur most commonly in warm water with a high bacterial count. *Naegleria* infections occur mostly in healthy young who acquire the protozoa into nasal passages from swimming, diving, and other water activity in warm freshwater lakes. **Primary amebic meningoencephalitis** causes massive brain and spinal tissue destruction when the organisms enter the CNS and has a greater than 95% mortality rate. *Acanthamoeba* enters by way of a skin cut, causing **granulomatous amebic encephalitis**. The amebae cause skin lesions that are followed months later by the onset of CNS disease, resulting in a high mortality rate. *Acanthamoeba* keratitis is a corneal infection associated with minor corneal trauma and the use of soft contact lenses. Treatment of this disease has been more successful than treatment of amebic encephalitides because delivery of high concentrations of antimicrobial drugs to infections or surgical debridement is possible.

The sexually transmitted disease **trichomoniasis** caused by the flagellete *Trichomoniasis vaginalis* is transmitted by contact because *T. vaginalis* does not form cysts. Contaminated water is the primary source of the cysts of *Giardia lamblia*, a food- and waterborne flagellate that causes gastrointestinal disease. *G. lamblia* is the most common cause of epidemic waterborne diarrheal disease in the United States. Some parasitic flagellates (**hemoflagellates**) grow inside blood cells and tissues, causing serious diseases. Blood-sucking insect vectors, serving as intermediate hosts in their complex life cycle, spread the flagellates. **Tsetse flies** spread **African sleeping sickness** caused by *Trypanosoma brucei*, a serious disease occurring in the African tropics. **Reduviid** ("kissing") **bugs** spread **Chaga's disease** caused by *Trypanosoma cruzi* occurring in tropical Latin America. Both of these trypanosomiases involve the lymphatic system and the CNS. Elimination of the insect vector is an effective control method. **Sand flies** spread **leishmaniasis** caused by *Leishmania*. The three types of leishmaniasis, in order of increasing severity, are **cutaneous leishmaniasis** (*L. tropica*), **mucocutaneous leishmaniasis** (*L. brasiliensis*), and **systemic leishmaniasis** (*L. donovani*). Systemic leishmaniasis develops into a deadly form (**kalazar**) with a mortality rate of 75% or higher in untreated cases.

Malaria, the major protozoan disease worldwide, is caused by *Plasmodium* species. The asexual phase of the life cycle occurs in the human and the sexual phase in the mosquito. Both hosts are required for its life cycle to be complete. Within hours after sporozoites have been injected from a mosquito bite, they invade liver cells where they undergo multiple asexual divisions (**schizogony**), and after 5 to 10 days packed liver cells burst and release thousands of merozoites into circulation. The merozoites attach to receptors on the red blood cell surfaces and invade red blood cells. Multiple divisions of the merozoites produce a red blood cell (**schizont**) with many more merozoites. The red blood cell bursts, releasing the merozoites to infect more red blood cells. Alternating bouts of chills, fever, and sweating occurring at regular intervals of 2 or 3 days result from synchronous red blood cells bursts. Certain merozoites differentiate into microgametes and macrogametes that are acquired by the mosquito. The sexual phase (**sporogony**) in the mosquito yields a diploid oocyst from the fertilization of the macrogamete by the microgamete. Sporozoites formed by the meiotic division of the diploid oocysts are then transferred by the bite of the infected female *Anopheles* mosquito to a human. The most serious form, falciparum malaria, can lead to complications of hemolytic anemia and damage the spleen, liver, and kidneys. Control is difficult due to the adaptive and survival capacity of the parasite and its vector. Some chemotherapeutic agents have limited control. Efforts to develop an effective vaccine are complicated by the antigenic changes from sexual reproduction and by the diversity of *Plasmodium* species.

Toxoplasmosis caused by *Toxoplasma gondii*, a widespread frequently asymptomatic human infection acquired from oocytes consumed with raw meat contaminated with animal feces, can be serious in immunocompromised individuals. Infection in pregnant women can lead to congenital infection during the first and second trimesters with fetal damage and possible abortion.

The emerging disease **cryptosporidiosis**, caused by *Cryptosporidium*, occurs where there is unacceptable water and food sanitation. Infection begins with ingestion of oocytes that infect intestinal cells, causing an unpleasant self-limiting gastrointestinal illness. Birds, fish, and mammals serve as reservoirs of *Cryptosporidium*. Infection by *Cyclospora cayetanensis*, another emerging disease, leads to watery diarrhea and general gastrointestinal distress that can persist for several weeks when untreated. Infection is acquired from ingestion of viable sporulated oocysts contaminating imported fruits and vegetables from southern countries (Mexico, Central America, etc.) or other food and water sources. Foodborne infection can be limited by simply washing fruits and vegetables before consuming them.

The harmful action of algae on humans is limited to a few species. The "red tide" involves the red-colored marine **dinoflagellates** *Gymnodium* and *Gonyaulax* that can produce algal blooms in warm waters during the summer months and release a potent muscle toxin. The toxin is concentrated in shellfish by their filter feeding. Human consumption of contaminated shellfish causes **paralytic shellfish poisoning**. Another toxin poisoning, called **ciguatera**, results from eating marine fish (grouper, snapper, etc.) that have consumed the dinoflagellate *Gambierdiscus toxicus*. This toxin is heat stable and causes gastrointestinal distress and CNS involvement.

The freshwater dinoflagellete *Pfiesteria piscicida*, recently discovered, releases a neurotoxin that causes neurologic damage and a necrotoxin producing bloody skin lesions in fish and humans who come in contact with water contaminated by the toxin. Reasons for these outbreaks are under investigation, but suspicion has been cast on pollution effluents from pig farms.

The only alga noted to cause infection is a green alga, *Prototheca moriformis*. Its habitat is soil. Human infection leads to **prothecosis**, a subcutaneous infection that can spread through the lymph system.

Topic Test 3: Diseases Caused by Protozoa and Algae

True/False

1. Antibiotics and chemotherapeutic agents most effectively control protozoal diseases.

2. The infection and the release of toxins are equally responsible for human diseases caused by algae.

3. *Giardia lamblia* is the most common cause of epidemic waterborne diarrheal disease in the United States.

Multiple Choice

4. Transmission of what protozoal form does not involve insect vectors and does not require direct contact?
   a. Cyst
   b. Sporozoite
   c. Microgamete
   d. Macrogamete

5. The form of the malarial parasite that infects red blood cells and is responsible for the reoccurring episodes of chills, fever, and sweating is the
   a. merozoite.
   b. oocyst.
   c. sporozoite.

Short Answer

6. Which growth form is present in the life cycle of all protozoa, and which protozoal growth form can be transmitted without requiring direct contact or an insect vector?

Topic Test 3: Answers

1. **False.** Because protozoa are eukaryotics, antibiotics and chemotherapeutic agents are limited in usefulness. The most effective control of protozoal disease is to control the route of transmission.

2. **False.** Among the algae that cause human disease, only one is noted to be associated with infections.

3. **True.** Although other protozoans induce waterborne diseases, fortunately they are less common.

4. **a.** The cyst form can persist in the environment.

5. **a.** Red blood cells have receptors to which merozoites can attach and subsequently invade.

6. The trophozoite is a form that is present in the life cycles of all protozoa. The cyst persists in the environment and is transmitted without the aid of an insect vector or direct contact.

Chapter Test

True/False

1. Most fungi are found parasitizing plants or decaying organic matter in the soil.

2. Species are distinguished among the fungi, except the yeast, by the type of reproductive structures they produce.

3. Superficial and cutaneous mycoses occur most frequently in humid situations.

4. Morphologic characteristics are more significant in identifying fungi than biochemical ones.

Multiple Choice

5. Which of the following represent ways of preventing fungal infections?
   a. Vaccination
   b. Chlorination of drinking water
   c. Daily consumption of vitamin A
   d. Wearing a mask when working with decaying vegetation or cleaning bird droppings
   e. Washing all vegetables to be consumed in bleach

6. The vegetative growth form of most fungi is
   a. long branched hyphae.
   b. chains of spheres.
   c. bricklike.
   d. submicroscopic.
   e. without a cell wall.

7. Why are fungal diseases, such as coccidioidomycoses or histoplasmosis, so widespread?
   a. Colonize decaying organic matter in many areas
   b. Produce lytic enzymes that digest tissue
   c. Produce virulent toxins
   d. Form airborne spores
   e. Can satisfy their iron requirement from human blood

8. What characteristics have proven best for classifying and identifying fungi?
   a. Fermentation reactions
   b. Cell wall components

    c. Reproductive structures
    d. Nitrogen sources
    .e. Vitamin requirements

9. Most fungal infections can be treated with
    a. penicillin.
    b. amphotericin B.
    c. AZT.
    d. isoniazid.
    e. streptomycin.

10. Which of the following is *not* present in protozoans?
    a. Cell wall
    b. Flagella
    c. Mitochondria
    d. Nucleus
    e. Ribosomes

11. In what human cells does the sexual stage of the malarial life cycle occur?
    a. Lymphocytes
    b. Macrophages
    c. Adipose cells
    d. Erythrocytes
    e. Hepatocytes

12. Amebic infections of humans are contracted primarily from
    a. mosquito bites.
    b. droplets from other humans.
    c. contaminated food and water.
    d. contaminated needles.
    e. None of the above

Short Answer

13. What is the best method currently to control the occurrence of malaria?

14. Fungal infections rarely are considered serious for humans unless they have AIDS, are transplant recipients, or have inherited immune dysfunctions. Why?

15. Why are algae not typically associated with infections?

Chapter Test Answers

    1. **True**

    2. **True**

    3. **True**

    4. **True**

    5. **d**

    6. **a**

7. **d**

8. **c**

9. **b**

10. **a**

11. **d**

12. **c**

13. Reduce populations of mosquitoes with insecticides and destroy their breeding sites.

14. Fungi are very sensitive to the cytotoxins released by phagocytic cells, T_C lymphocytes, and T_{NK} cells. Also, fungi do not compete successfully with bacteria for space or nutrients on body surfaces.

15. Animal tissue is a nonhospitable environment for algal growth.

Check Your Performance

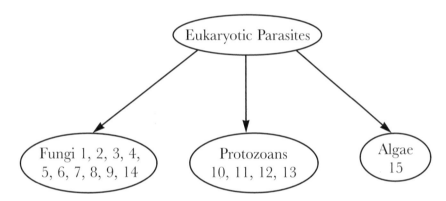

Note the number of questions in each grouping that you got wrong on the chapter test. Identify areas where you need further review and go back to relevant parts of this chapter.

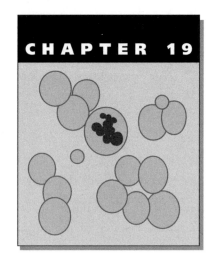

CHAPTER 19

Environmental Microbiology

Microorganisms play an important role in the environment. Some of the things they do are beneficial, such as breaking down herbicides and pesticides to nontoxic chemicals. Other things that microorganisms do can be extremely harmful, such as metabolizing elemental sulfur to sulfuric acid.

ESSENTIAL BACKGROUND

- **General chemistry (Chapter 2)**

TOPIC 1: BIOGEOCHEMICAL CYCLES: CARBON, NITROGEN, SULFUR, AND IRON CYCLES

KEY POINTS

✓ *How do microorganisms fix carbon dioxide, nitrogen, sulfur and iron?*

Carbon is constantly cycling between complex **organic** molecules, such as amino acids, sugars, and lipids, and **inorganic** forms, such as carbon dioxide and carbon monoxide.

Complex organic molecules are broken down by metabolism to carbon dioxide. Carbon dioxide or carbon monoxide can be used as hydrogen acceptors to form methane gas (CH_4). The carbon dioxide or carbon monoxide can also be fixed or converted to complex organic molecules by the Calvin cycle or other metabolic pathways. This constant cycling between complex organic molecules and carbon dioxide is called the **carbon cycle**.

Carbon dioxide comes into the carbon cycle from the respiration process of animals and many bacteria. Combustion adds carbon dioxide to the atmosphere. Another source of inorganic carbon is the erosion of carbon-containing minerals.

Some bacteria, called **autotrophs**, and plants use the carbon dioxide and build it back up into complex organic molecules. Autotrophic bacteria may be **photoautotrophic**, such as the green and purple bacteria and *Cyanobacteria*. The photoautotrophs build complex carbon molecules through photosynthesis. Other autotrophic bacteria are **chemoautotrophs or chemolithotrophs**. *Thiobacillus* species are chemoautotrophs. Chemoautotrophs oxidize minerals for energy rather than using photosynthesis.

Autotrophic bacteria use carbon dioxide as their sole carbon and energy source. These organisms benefit the environment by using the carbon dioxide that holds heat near the surface of the earth—the "greenhouse effect." Some bacteria are able to metabolize carbon monoxide into usable substrates.

Autotrophic bacteria also start the carbon up the food chain. They fix carbon dioxide into organic molecules for their own use. The bacteria are often food for protozoans, which become food for higher animals and so the complex carbon molecules work up the food chain. The primary pathway for **fixing carbon dioxide** into organic molecules is the Calvin cycle.

Nitrogen is another element that cycles through the environment from elemental form through **ammonia** to **nitrite** and **nitrate**. Some nitrogen gets into the soil from synthetic nitrogen fertilizers. Nitrogen may also get into the soil from plants. Much of the nitrogen in the soil comes from bacteria that **fix elemental nitrogen** into a form that plants can use, such as ammonia, nitrite, and nitrate.

The processes in the nitrogen cycle are as follows:

1. Nitrogen fixation: N_2 elemental nitrogen $\rightarrow NH_3$ ammonia

2. Oxidation of ammonia: $NH_3 \rightarrow NO_2^-$ Nitrite $\rightarrow NO_3$ Nitrate

3. Nitrate reduction: NO_3 Nitrate $\rightarrow NO_2^-$ nitrite

4. Ammonification: NO_2^- Nitrite $\rightarrow NH_3$ ammonia

5. Denitrification: $NO_2^- \rightarrow NO$ Nitric oxide $\rightarrow N_2O$ nitrous oxide $\rightarrow N_2$ elemental nitrogen that can be cycled back in via nitrogen fixation.

Bacteria that fix elemental nitrogen into ammonia include species of *Chromatium, Rhodopseudomonas, Klebsiella, Azospirillum, Bacillus, Clostridium, Azotobacter,* Cyanobacterium, and the root nodule bacteria *Rhizobium. Nitrosomonas, Nitrospora,* and *Nitrococcus* oxidize ammonia to nitrate. *Nitrobacter* oxidizes nitrite to nitrate.

Denitrifying bacteria convert nitrate to elemental nitrogen and return it to the atmosphere. Some denitrifying bacteria are species of *Pseudomonas, Rhizobium, Rhodopseudomonas,* and *Propionibacterium.*

Nitrate-reducing bacteria are those that reduce nitrate to nitrite. Some of the organisms that are nitrate reducers are species of *Escherichia, Nocardia, Staphylococcus,* and *Enterobacter.* Other organisms may reduce nitrate under anaerobic conditions. They use the nitrate as a terminal electron acceptor in energy generation.

Sulfur is found in deposits in the earth's crust. It is also found with coal deposits. There are many **soluble salts** of sulfur found in bodies of water. Car exhaust adds sulfur compounds to the air. Sulfur is found in cells, generally in the **sulfhydryl groups** (−SH) of amino acids, such as cysteine.

Most **heterotrophic** bacteria take up sulfur as a sulfate salt. Some bacteria can use **sulfides**, such as hydrogen sulfide. Sulfides are toxic to many organisms. The organisms that use sulfides react serine with hydrogen sulfide (H_2S) to form cysteine.

Organisms such as *Beggiatoa* oxidize hydrogen sulfide to elemental sulfur. Other organisms, such as *Thiobacillus,* are **acidophilic** (acid loving) and oxidize hydrogen sulfide to sulfuric acid (H_2SO_4).

The processes of the sulfur cycle include reactions with iron in several steps:

1. Sulfur reduction: S elemental sulfur → H_2S hydrogen sulfide

2. Hydrogen sulfide oxidation: H_2S hydrogen sulfide → S elemental sulfur

3. Sulfur oxidation: S elemental sulfur react with water → H_2SO_4 sulfuric acid

4. Dissimilatory sulfate: H_2SO_4 sulfuric acid → H_2S hydrogen sulfide and oxygen

5. Desulfuration: H_2S hydrogen sulfide react with ferrous iron → FeS ferrous sulfide and hydrogen

6. Sulfuric acid reacting with iron: H_2SO_4 | Fe → $Fe(OH)_2$ ferrous hydroxide + FeS ferrous sulfide + hydroxy ions OH^-

7. Sulfuric acid assimilation: H_2SO_4 + organic compounds → organic sulfur compounds, such as the amino acid cysteine.

Iron normally cycles between the **ferric** or +3 state and the **ferrous** or +2 state in nature. Iron is, of course, mined for the production of steel. It is found in many mineral salts, such as ferricyanide salts. Iron is also found in cells, usually as a **hemin chelate** in the cytochromes of cells (also the hemoglobin and myoglobin in animals).

Some iron salts are soluble. Others, such as ferric hydroxide, precipitate under alkaline conditions, returning them to the mineral base of the earth's crust.

Topic Test 1: Biogeochemical Cycles: Carbon, Nitrogen, Sulfur, and Iron Cycles

True/False

1. Carbon monoxide is toxic to all microorganisms.

2. Carbon dioxide is fixed by the TCA cycle.

3. Acidophiles, such as *Thiobacillus* digest sulfuric acid.

Multiple Choice

4. Elemental nitrogen is fixed into ammonia by
   a. *Thiobacillus.*
   b. *Clostridium.*
   c. *Rhizobium.*
   d. b and c

5. Sulfur exists in cells as
   a. digestive acids.
   b. as a chelated mineral in cytochromes.
   c. as sulfhydryl groups of amino acids.

Short Answer

6. List the steps of denitrification.

Topic Test 1: Answers

1. **False.** Some microorganisms can use carbon monoxide.

2. **False.** Carbon dioxide is fixed by the Calvin cycle.

3. **False.** Acidophiles, such as *Thiobacillus*, produce sulfuric acid.

4. **d**

5. **c**

6. In denitrification, nitrate goes to nitrite, to nitric oxide, to nitrous oxide, and finally to elemental nitrogen.

TOPIC 2: EXTREMOPHILES: THEIR OCCURRENCE AND ACTIVITIES IN THE ENVIRONMENT

KEY POINTS

✓ *What are extremophiles?*

✓ *Where are extremophiles found?*

Extremophiles are organisms that thrive in **environmental extremes**. **Table 19.1** summarizes the extreme conditions, where they may be found, and lists examples of the organisms that fall into the extreme categories.

Topic Test 2: Extremophiles: Their Occurrence and Activities in the Environment

True/False

1. Halophiles like high NaCl or potassium concentrations.

2. Acidophiles are often also halophiles.

3. Nothing grows in the cooling water of nuclear reactors.

Multiple Choice

4. *Thiobacillus* is an example of a
   a. barophile.
   b. alkaliphile.
   c. acidophile.
   d. magnetotaxic organism.

Short Answer

5. Where might osmophiles be found?

		EXAMPLE ENVIRONMENT	
GROUP	CONDITION THE GROUP LIKES	LOCATION	EXAMPLE ORGANISMS
Obligae anaerobes	No oxygen	Bovine rumen	*Bacteroides ruminicola*
Temperature extremes			
Psychrophiles	Extreme cold down to −14°C	Polar sea	*Vibrio marinus*
Thermophiles	Extreme heat up to 105°C	Hot springs in Yellowstone Park	*Pyrococcus woesei*
Radiation tolerant	Survive gamma radiation	Nuclear reactor cooling water	*Dienococcus radiodurans*
Pressure extremes			
Barotolerant and barophilic	High pressure up to 1000 atmospheres	Bottom of the ocean	*Spirillum* species
Osmophilic or osmotolerant	High osmotic pressure	Jams and jellies	Molds: *Penicillium, Aspergillus* Yeast: *Zygosaccharomyes rouxii*
Halophilic	High salt Na and K conc. High potassium concentrations	Salt water	*Halobacter salinarium* *Halobacterium* species
Halotolerant	Tolerate high salt up to 7.5%	Salt water	*Staphylococcus aureus*
Xerotolerant	Low water activity	California desert	*Nostoc commune*
pH extremes			
Acidophiles	Low pH (as low as 1.0)	Acid mine runoff	*Bacillus acidocaldarius* species of *Halobacterium, Natronobacterium, Clostridium*
Alkaliphiles	High pH (may also be halotolerant)		
Oxidation-reduction potential (redox potential)	Donates electrons = high redox Receives electrons = low redox	Oxygenated environments Anaerobic environments	Aerobic bacteria Anaerobic bacteria: *Bacteroides* species
Magnetotaxic	Magnetic fields		*Aquaspirillum magnetotacticum*

Table 19.1. Extremophiles

Topic Test 2: Answers

1. **True.** Halophiles like high NaCl or potassium concentrations.

2. **False.** Alkaliphiles may be halotolerant, but acidophiles are not.

3. **False.** Radiation-tolerant organisms, such as *Dienococcus radiodurans*, grow in the cooling water of nuclear reactors.

4. **c**

5. Osmophiles can be found growing in jams and jellies that have high osmotic pressure.

TOPIC 3: HARMFUL AND BENEFICIAL ACTIONS OF MICROORGANISMS IN THE ENVIRONMENT

KEY POINTS

✓ *How do microorganisms hurt the environment?*

✓ *How do microorganisms help the environment?*

Microorganisms have diverse influences on the environment. Some of those influences are harmful and others are helpful. Among the **harmful** things microorganisms do to the environment is cause plant diseases. When diseased trees die, they become kindling for a forest fire. When plants die off, the soil may also erode. Organisms such as *Thiobacillus* species oxidize the sulfur found in mine tailings. The sulfur is oxidized to sulfuric acid. The **sulfuric acid** can damage fish and wildlife habitats.

Some microorganisms actually increase the toxicity of substances. The organisms may add a methyl-group to mercury, arsenic, and selenium, which increases the toxicity of these metals and metalloids. The process of microbes increasing toxicity of heavy metals is a form of **biomagnification**. *Desulfovibrio desulfuricans* is one organism that can methylate mercury.

Microorganisms are responsible for introducing many toxic substances into the food chain. The microbes take up and chelate heavy metals. The **chelation** accumulates the metals in the cells. When the microorganisms are used as food by other organisms, the heavy metals begin their journey up the food chain. The process is the same for naturally occurring heavy metals and heavy metal pollutants.

Microorganisms bring other toxins into the **food chain**. Lipid-soluble pollutants, such as polychlorinated biphenyls, DDT, and others, are taken up by the microbes and passed up the food chain when the microbes become food for other organisms.

Some actions of microorganisms in the environment can be both beneficial and destructive, depending on the circumstances. An example of this is the breakdown of crude oil and petroleum products by microorganisms. It can be good for cleaning up oil spills when there is a large surface area for microbial activity and the gasses from the degradation can disperse into the atmosphere. The process can also be destructive when organisms such as *Methylomonas* species **degrade hydrocarbons** in the soil. The degradation uses up the oxygen near the roots of plants and produces methane and other toxic gases from the hydrocarbons. Hydrogen sulfide may also be produced in the degradation of hydrocarbons. The degradation products can damage plants and trees.

Beneficial actions of microorganisms in nature include fixing atmospheric nitrogen for plant use. The organisms that do the **nitrogen fixation** may live in root nodules of leguminous plants or may be free living. This is extremely important for agriculture. These organisms and the nitrogen fixation process are discussed in Topic 1 of this chapter.

Another important action of microorganisms in the environment is the degradation of plant matter that accumulates on the floor of the forest and elsewhere. By degrading the materials, the organisms release the nutrients back to the soil. The degradation of dead plant material also removes it as a potential fuel for a forest fire.

Microorganisms degrade complex organic compounds, such as those found in **sewage**. Many organisms are used by waste treatment plants to break down the waste.

Sewage is water that contains human or industrial waste; used water, such as that used for washing; grease; and anything else people decide to dump down the drain. There are three steps in the sewage treatment process:

1. **Solids are removed** from the sewage. Some of the solids are filtered out, others precipitated, and polar substances such as grease may be removed by skimming. The effluent from this step proceeds to step 2. The sludge may go into an anaerobic digester.

2. **Removal of organic matter**. This can be accomplished by the **trickling filter method**, where the effluent percolates through a rock bed that is aerated and coated with aerobic bacteria. An alternative to the trickling filter is the "**activated sludge**" **method**. The method agitates and aerates the effluent in a processing tank. Sheathed bacteria, such as *Sphaerotilus* and *Beggiatoa*, are usually found on the rocks in the trickling filter and in the tanks used for the activated sludge treatment.

3. **Final filtration and chemical treatment** of the effluent from step 2. At this step, the effluent may be filtered through fine sand or charcoal or both. Two types of chemical treatment are done at this step. The first type of treatment is the removal of **phosphates**. If the water contains too many phosphates, it can stimulate the growth of algae in rivers and lakes. The second type of treatment is **chemical disinfection** of the water. This is usually done with **chlorine**, either as a gas or as hypochlorite. The water can also be disinfected with **ozone**. The disadvantages to ozone treatment are that it is expensive and there is no residual to help keep the water disinfected. Once the water is disinfected, it can be recycled into the municipal water system or can be discharged into the environment. The sludge from both steps 1 and 2 goes into **sludge digesters**. These are tanks where anaerobic bacteria break the organic matter down into simple organic molecules, carbon dioxide and methane (CH_4) gas. The methane from the digester can be used to heat the digesters or for other industrial uses.

Some microorganisms **degrade industrial chemicals**, such as some herbicides and pesticides. The process of using microorganisms to remove pollutants from the environment is called **bioremediation**. Some of the chemicals that microorganisms can break down are benzene, toluene, xylene, creosote, fuel oil, and trichloroethylene. A fungus, *Phanerochaete chrysosporium*, is able to attack the pesticide DDT. Halocarbons, such as trichloromethane, chloroform, and others have been degraded by *Methylococcus* species.

Microorganisms have not been as successful in breaking down many plastic polymers such as polyvinyl chloride, polystyrene, and polyethylene. These plastics are **biologically inert** to the extent that they are used for blood and tissue contact medical instruments. Because they are biologically inert, they are not poisoning the environment. Some work is focused on finding microorganisms that can degrade these materials just to rid the landfills of them.

Many bacteria have some effect on **ice nucleation**—the formation of ice crystals. Some bacteria cause nucleation, which can damage plants. Other bacteria inhibit nucleation. Some of the nucleation inhibitory bacteria are being used agriculturally to protect plants from freezing and frost damage. *Pseudomonas fluorescens* has been licensed for use as a frost protectant for agriculture. These are just a few examples of the actions of microorganisms in the environment.

Topic Test 3: Harmful and Beneficial Actions of Microorganisms in the Environment

True/False

1. Nothing can break down the pesticide DDT.

2. Bacteria can cause ice crystals to form on plants.

3. Degradation of hydrocarbons can damage plant roots.

4. Aerobic bacteria in trickling filters break down organic matter in sewage treatment.

Multiple Choice

5. Sulfuric acid is produced from mine tailings by
   a. *Thiobacillus.*
   b. *Pseudomonas.*
   c. *Bacteroides.*

6. Some of the industrial chemicals that can be degraded by microorganisms are
   a. toluene.
   b. benzene.
   c. trichloroethylene.
   d. polystyrene.
   e. a, b, and c

Short Answer

7. How do microorganisms help toxic chemicals get into the food chain?

Topic Test 3: Answers

1. **False.** A fungus, *Phanerochaete chrysosporium*, can break down the pesticide DDT.

2. **True**

3. **True**

4. **True**

5. **a**

6. **e**

7. Toxic chemicals get into the food chain when they are taken up by bacteria. Heavy metals are chelated into structures in the bacterial cells and accumulate that way. Lipid-soluble chemicals get into lipid-containing structures. The microorganisms are used for food by other organisms and pass the toxins up the food chain.

APPLICATION

The alkaliphilic bacteria *Bacillus alcalophilus* and *Bacillus pasteurii* produce enzymes that are used in laundry detergents. The enzymes remove grease and protein stains. The enzymes are functional in highly alkaline conditions and in the presence of detergents.

DEMONSTRATION PROBLEM

If you wanted to isolate and study some bacteria that facilitate ice nucleation, where should you look for them?

Solution

A good place to find these organisms is in a snowball. The organisms trapped in the ice crystals that make up the snowflakes are likely to be good ice nucleators.

Chapter Test

True/False

1. Microorganisms contribute to water damage of steel and iron structures.

2. Chemoautotrophic organisms are also called chemolithotrophic organisms.

3. Iron cycles between the ferrous and ferric states in the environment and in cells.

4. Bacteria can make heavy metals more toxic by adding a methyl group to the metals.

5. Bacteria accumulate heavy metals by chelating them.

Multiple Choice

6. Iron in cells is commonly found in
   a. heme.
   b. cytochromes.
   c. cell proteins.
   d. a and b

7. Sulfur compounds combine with _____ to form cysteine in bacterial cells.
   a. cytosine
   b. serine
   c. adenine
   d. glutamine

8. Nitrogen fixing bacteria can be found
   a. free-living in soil.
   b. in root nodules of leguminous plants.
   c. in root nodules of deciduous plants.
   d. a and b

9. When organisms make toxic substances more toxic, the process is called
   a. bioremediation.
   b. biomagnification.
   c. biotoxification.

10. When organisms break down toxic materials in the environment, the process is called
    a. bioremediation.
    b. biomagnification.
    c. biodetoxification.

11. Extreme thermophiles are found
    a. in polar seas.
    b. in hot springs.
    c. around thermal vents on the ocean floor.

Short Answer

12. Why should we be concerned about the presence of bacteria that cause ice nucleation?
13. What is magnetotaxis?
14. What kinds of organisms use photosynthesis?
15. What do chemoautotrophic organisms use for energy?
16. What does the name acidophile mean?

Chapter Test Answers

1. **True**
2. **True**
3. **True**
4. **True**
5. **True**
6. **d**
7. **b**
8. **d**
9. **b**
10. **a**
11. **c**
12. Ice crystals cause frost damage to plants. If there are bacteria present that stimulate ice nucleation, they help damage plants.
13. Magnetotaxis is the attraction of some microorganisms to magnetic fields.
14. Photoautotrophic bacteria and plants are photosynthetic organisms.
15. Chemoautotrophic organisms use inorganic substances for energy.
16. Acidophile means acid loving.

Check Your Performance

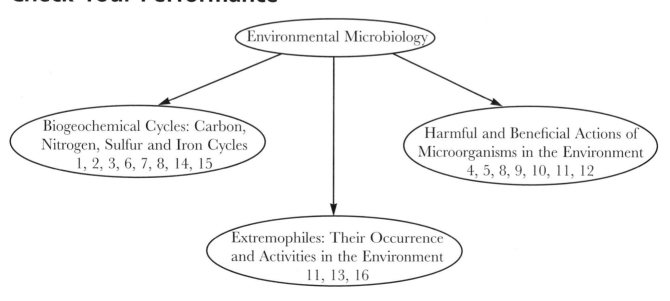

Environmental Microbiology

Biogeochemical Cycles: Carbon, Nitrogen, Sulfur and Iron Cycles
1, 2, 3, 6, 7, 8, 14, 15

Harmful and Beneficial Actions of Microorganisms in the Environment
4, 5, 8, 9, 10, 11, 12

Extremophiles: Their Occurrence and Activities in the Environment
11, 13, 16

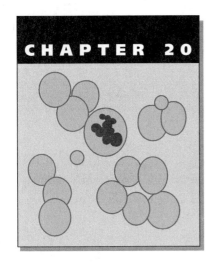

CHAPTER 20

Biotechnology and Applied Microbiology

Humankind has wanted to maintain favorable combinations of inheritance, particularly in domesticated animals and plants, that is, to **clone** individuals with these traits. To do this has been difficult at best, and impossible in many instances to do. Many dreamed of the time when desirable inherited traits could be fused deliberately and incorporated into progeny on a large scale. One would assemble **genes** the way an engineer develops parts to be assembled into a new machine, hence coining the term **genetic engineering**.

Genetic engineering began to take form experimentally in the early 1970s because four factors were brought together: knowledge of the biochemistry of DNA, understanding of how DNA directs protein synthesis, discovery of enzymes permitting cutting and splicing of DNA, and development of suitable hosts in which to clone DNA fragments. The engineered DNA (**recombinant DNA**) results from the fusion of DNA fragments from different sources. The first suitable hosts were bacteria. Only bacterial hosts for cloning recombinant DNA are considered here.

ESSENTIAL BACKGROUND

- **Structure of DNA (Chapter 2, Topic 3)**
- **Replication, transcription, translation in prokaryotes (Chapter 10, Topic 1)**
- **Plasmids (Chapter 11, Topic 1)**
- **Transformation (Chapter 11, Topic 4)**

TOPIC 1: RECOMBINANT DNA METHODOLOGY

KEY POINTS

✓ *How is recombinant DNA constructed?*

✓ *What is the significance of a vector when working with recombinant DNA?*

✓ *What is a suitable host cell?*

✓ *Why are marker genes significant when probing a genomic library?*

DNA sequences are spliced together from different species and/or even artificially constructed DNA sequences to form **recombinant DNA**. One of the DNA fragments spliced into the recombinant DNA is a gene of interest from one organism that one wishes to introduce into a second organism, for example, the introduction of human insulin gene into the bacterium *Escherichia coli* or the gene for insect toxin of *Bacillus thuringensis* into corn plants.

First, one must extract and purify the DNA from the organism having the desired trait. If one is not certain where the gene of interest from this organism is located in the genome, then the entire genome (all of the organism's chromosomes) must be used. Sometimes one knows the location of the gene of interest and can excise and separate that gene from the rest of the genome. If one is going to place the gene from a eukaryotic organism, for example an immunoglobulin gene (Chapter 14, Topic 5), into a prokaryotic one, there is a problem of the introns in eukaryotic genes. Prokaryotes, such as bacteria, cannot recognize exons from introns nor do they have the enzymatic machinery to remove introns. By isolating mRNA from cells with the desired gene functioning and using the enzyme reverse transcriptase (RNA dependent DNA polymerase) from retroviruses, one can obtain a DNA sequence without introns. To obtain enough DNA for the subsequent recombinant DNA processes, one may clone the DNA sequence obtained from RNA by using **polymerase chain reaction**.

Second, **vector** DNA is needed to transmit the recombinant gene into host cells. Often the vector is a plasmid, although viruses, microchromosomes found in yeast, or carborundum particles coated with DNA for transforming plants may be vectors. If a plasmid is used, it is constructed of DNA fragments from several sources. Some fragments may be synthetic sequences designed to provide sites for opening circular plasmids and inserting recombinant DNA. Vectors should have the following characteristics:

1. an origin of replication to permit replication in the host cell;

2. a marker gene, usually for antibiotic resistance such as ampicillin, to select for host cells containing plasmids;

3. a promoter and operator functional in the host cell to permit proteins to be made using the information in the recombinant gene;

4. a marker gene that is altered in function or "knocked out" when there is recombinant DNA in the plasmid;

5. a polylinker region into which to insert the recombinant DNA.

Third, **restriction endonucleases** from various bacteria cut DNA molecules at very specific sequences. The resulting DNA fragment may be blunt ended or may have single-stranded "tails" at each end of the cut because the latter restriction endonucleases do not hydrolyze the DNA strands at exactly opposite points but rather offset by two to four bases. Usually, restriction endonucleases producing tails are used. When DNA fragments are spliced together, splicing occurs more readily when complementary tails from different DNA fragments hydrogen bond with one another. The polylinker in the vector permits potential use of many different restriction endonucleases to open the plasmid, depending on which restriction endonuclease is used to cut the DNA to be recombined. Both the vector DNA and the source DNA are digested with the same restriction endonuclease. Sometimes two different restriction endonucleases are used so that the tails on the DNA fragments are different at each end, permitting splicing between the plasmid and the source DNA to occur in only one orientation.

Fourth, DNA fragments are spliced into the vector plasmid. A DNA ligase accomplishes this if ATP is provided to drive the reaction. The DNAs to be ligated are mixed together with DNA ligase and ATP and then allowed to incubate quietly in the cold. The lack of agitation and cold favor hydrogen bonding of the single-strand DNA tails of the DNA fragments. T_4 DNA ligase, coded in a gene of T_4 virus and synthesized in T_4-infected *E. coli*, is used usually because it readily splices gaps in DNA strands whenever it encounters them and is relatively stable compared with many other DNA ligases.

Fifth, recombinant DNA must be cloned and its information converted into proteins. A suitable host is required and must be transformed with the recombinant plasmid. Because we are concentrating on prokaryotic situations, a suitable bacterium is needed. The host should have the following characteristics:

1. **Relaxed** host so that many plasmids (up to 20) can replicate within a host cell; the opposite is a **stringent** host where less than three plasmids are supported;

2. Susceptible to the antibiotic for which the plasmid carries a resistance gene to select against hosts without plasmids;

3. Lack the marker gene used in the plasmid to determine recombination;

4. Have mutations that alter membrane permeability under environmental stress so plamids may be introduced into host cells.

"Heat shock" is a mechanism for triggering transformation by plasmids into bacteria. Under good conditions, one bacterium of every thousand will contain a plasmid. Heat shocking involves chilling the bacteria in ice, plunging them into a 42°C water bath for 90 seconds, and then chilling them on ice again. No fancy equipment is used. If one has an electroporator that jolts the bacteria with a quick electrical pulse, one in every 100 bacteria will absorb a plasmid.

Finally, one must be able to recognize which host cells have the recombinant DNA of interest. The medium used to culture the cells contains the antibiotic for which the plasmid gene provides resistance and the reagents necessary to recognize cells carrying recombinant plasmids from those having nonrecombinant plasmids. A second selection is necessary to recognize host cells carrying plasmids with the recombinant gene of interest. Occasionally, the genes of interest produce readily observable physiologic responses, such as a fluorescent pigment or bioluminescence. Usually, this is not the case, and some type of **probe** must be used to see which recombinant cells have the recombinant DNA of interest. Recombinant colonies are cultured one to a well of a multiwell plate. A replica of the colonies in the wells is made on a membrane and each colony probed. Three types of probes can be used: RNA (**Northern blot**), DNA (**Southern blot**), or protein (**Western blot**). If one knows a 20 to 40 base sequence of the gene of interest, one can construct a complementary sequence of either RNA or DNA and probe for mRNA or the gene itself. The probes are labeled radioactively, fluorescently, bioluminescently, or enzyme linked. The protein produced from the gene of interest may be probed by immunofluorescent antibody assay or enzyme-linked assay using appropriately tagged monoclonal antibody to the protein. The colonies of host cells having recombinant plasmids are said to make up a **genomic library** of the DNA of the species providing the recombinant genes. The problem with this "library" is that the "books" have no titles on them.

Topic Test 1: Recombinant DNA Methodology

True/False

1. Recombinant DNA refers to reshuffling genes during meiosis in a sexually reproducing organism.

2. In biotechnology the term vector is used to designate DNA that carries a recombinant gene.

Multiple Choice

3. Processes used to manipulate DNA when forming recombinant DNA have become known as
   a. advanced recombination.
   b. synthetic DNA
   c. genetic engineering.
   d. manipulated DNA.
   e. exploitative genetics.

4. Suitable host cells require which of the following characteristics?
   a. Relaxed toward plasmid replication
   b. Mutations in membrane permeability
   c. Susceptibility to an antibiotic, such as ampicillin
   d. Lack of some marker gene present on the vector DNA
   e. All of the above
   f. None of the above

5. Cells shown to contain recombinant plasmids are said to make a(n)
   a. genomic library.
   b. mutated cell population.
   c. exploitatively manipulated cell.
   d. restriction endonuclease.
   e. infectored cell.
   f. None of the above

Short Answer

6. What are the six processes necessary to recombine and clone DNA?

Topic Test 1: Answers

1. **False.** Recombinant DNA is genetic material from different sources (species) that has been spliced together deliberately.

2. **True.** The analogy is to the insect that transmits a virus from one mammalian host to another. Here a gene is being transferred from one species to another.

3. **c.** The construction of recombinant DNA is said to resemble the construction of a machine.

4. **e.** a is to ensure cloning of the vector, b to permit uptake of the vector, c to select for cells containing a plasmid, and d to distinguish host cells containing recombinant plasmids.

5. **a.** Cells carrying different recombinant plasmids are considered like the many books found in a library, only they lack titles so one must hunt to find the desired "book."

6. Extract and purify source DNA, obtain DNA of suitable vector, digest source DNA and vector DNA with the same restriction endonuclease(s), use T_4 DNA ligase to splice source DNA into vector DNA, introduce recombinant DNA into suitable host cells, and use marker traits and/or probes to select the population of host cells harboring the desired recombinant DNA.

TOPIC 2: APPLICATIONS: AGRICULTURE, ENERGY PRODUCTION, AND MEDICINE

KEY POINTS

✓ *How has recombinant DNA been exploited?*

✓ *What existing or potential benefits have arisen or could arise by using biotechnology?*

Recombinant DNA (rDNA) technology has been used to genetically alter *E. coli* to benefit our health and welfare. **Metabolic pathway engineering** of *E. coli* is another application of rDNA technology. Instead of the rDNA gene protein product being the goal, the rDNA proteins are designed to amplify or alter existing biosynthetic pathways for the overexpression of existing products or the production of new products.

One interesting example of metabolic engineering is the production of ethanol by *E. coli*. The potential of this to solve problems of the greenhouse effect from fossil fuel use and our dependence on foreign oil was recognized by the U.S. Patent Office, which awarded patent number 5,000,000 (round number was planned) for developing metabolically engineered *E. coli* with increased production of ethanol to replace fossil fuel as an energy source. Solar energy is stored in green plants by the conversion of carbon dioxide to biomass, most of which is lignocellulose containing the structural polymers cellulose, hemicellulose, pectin, and lignin. Eighty percent of lignocellulose consists of polymers with carbohydrate subunits used by *E. coli* as a source of carbon and energy. The remaining 20% contains lignin and other nonfermentable compounds. Developing commercially feasible depolymerization of the polymers remains a challenge.

Zymomonas mobilis bacteria have a native homo-ethanol fermentative pathway in which the end products are carbon dioxide and ethanol:

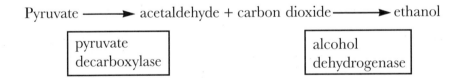

Pyruvate ⟶ acetaldehyde + carbon dioxide ⟶ ethanol

| pyruvate decarboxylase | alcohol dehydrogenase |

This pathway produces a higher yield of ethanol from fermentative growth on carbohydrates than any of the native fermentative pathways of *E. coli*, but *Zymomonas* lacks the ability to ferment pentose sugars. *E. coli* can grow fermentatively on pentose sugars that represent a large fraction of sugars in lignocellulose.

The fermentative pathways of metabolically engineered *E. coli* growing on carbohydrates all use pyruvate as their substrate. The K_m of pyruvate decarboxylase (PDC), the first enzyme of the homo-ethanol pathway cloned from *Z. mobilis*, is significantly less than the K_m's for either of the two native fermentative pathways of *E. coli*, greatly favoring homo-ethanol fermentation. Genetically improved *E. coli* for ethanol production by L. O. Ingram and colleagues has been achieved by chromosomal integration of the *Z. mobilis* genes immediately downstream from promoters (Chapter 10) of the pyruvate formate lyase gene. This has resulted in high expression efficiency of the PDC and alcohol dehydrogenase genes, destruction of a competing fermentative pathway, and 100% stability of the integrated genes compared with less stability when the genes are on plasmids, that may be spontaneously lost (Chapter 11).

Genetic engineering applications in agriculture have led to increased production efficiency of crop plants. One example is increasing the efficiency of nitrogen utilization by plants. The unicellular green alga, *Chlorella sorokiniana*, is a microorganism that lives in competition with other aquatic organisms for nutrients. The alga has evolved a highly efficient nitrogen utilization pathway possessing a reduced nicotinamide adenine dinucleotide phosphate-specific glutamate dehydrogenase enzyme not found in plants. Nitrogen fertilizer is the most costly and often limiting nutrient to agriculture in the United States and in particular to Third World countries. Moreover, use of large amounts of nitrogen fertilizer in the United States often results in run-off and contamination of freshwater bodies. In part by using the strategies outlined in the previous topic, R. R. Schmidt and staff at Monsanto Co. have cloned the *Chlorella* nitrogen-utilization genes into certain agronomically important plants. The process has been patented and results have been dramatic.

Biotechnology and rDNA technology have led to production of pharmaceutical proteins, insulin, human growth hormone, alpha interferon, and bovine growth hormone by genetically engineered bacteria. *E. coli* is most commonly chosen as a commercial source of cloned gene end products because of the existing broad and deep knowledge of its genetics and physiology, its rapid and easy growth, its asexual reproduction and haploid chromosome, and the desired number of copies of cloned genes is controllable. Similar to recent advances in information technology that have dominated the last years of the 20th century, biotechnology and rDNA technology are poised to have a similar impact in the early part of the 21st century.

Topic Test 2: Applications: Agriculture, Energy Production, and Medicine

True/False

1. rDNA is an essential component for achieving metabolic engineering of an organism.

2. Burning ethanol produced by metabolically engineered bacteria may replace fossil fuels.

Multiple Choice

3. Addition of genes from the alga *Chlorella sorokiana* into crop plants has significantly increased crop yield because of more efficient utilization of
   a. iron.
   b. nitrogen.
   c. phosphorus.
   d. sulfur.

4. Metabolic engineering of *E. coli* as a commercial source of fuel ethanol includes altering its
   a. carbohydrate catabolic pathways.
   b. fermentative pathways.

    c. TCA cycle.

    d. ability to grow autotrophically.

5. Which of the following components of lignocellulose is *not* susceptible to degradation and subsequent fermentation by metabolically engineered *E. coli* used for fuel ethanol production?

    a. Cellulose

    b. Hemicellulose

    c. Lignin

    d. Pectin

Short Answer

6. Briefly discuss why *E. coli* is commonly used as the organism in production of gene products by rDNA technology.

Topic Test 2: Answers

1. **True.** A critical early step in metabolic engineering is to obtain the needed genes, which requires rDNA.

2. **True.** Only fossil fuel use contributes to a net increase in carbon dioxide.

3. **b.** The gene introduced into plants codes for an enzyme that is highly efficient in nitrogen assimilation.

4. **b.** Manipulation of the fermentative pathways results in a high percent yield of ethanol.

5. **c.** Cellulose, hemicellulose, and pectin are polymers consisting of monomeric carbohydrate subunits, all of which are metabolizable by *E. coli* under fermentative conditions.

6. *E. coli* offers the advantages of being well understood regarding its genetic and physiological properties, able to easily grow and growing rapidly, a haploid organism that reproduces asexually, and susceptible to regulation of the number of copies of the cloned gene.

Chapter Test

True/False

1. Recombinant DNA is the fusion of the DNA from one organism with a protein from another organism.

2. The process of putting the genes for alcohol fermentation from *Z. mobilis* into the genome of *E. coli* is known as genetic engineering.

3. Polymerase chain reaction is a process used to splice strings of desirable genes together.

4. Restriction endonucleases hydrolyze DNA into fragments at very specific DNA sequences.

Multiple Choice

5. Which of the following conditions would *not* be found in a host cell suitable for cloning a plasmid?
   a. Relaxed host
   b. Antibiotic sensitive
   c. Temperature sensitive
   d. Increased membrane permeability under heat stress
   e. Lack of a marker gene

6. Use of what product would be reduced or eliminated by introducing glutamate genes of an alga into a corn plant?
   a. Herbicides
   b. Nitrogen fertilizer
   c. Potassium and calcium supplements
   d. Insecticides
   e. None of the above

7. Which of the following is *not* a factor for selecting *E. coli* as a host for cloning genes?
   a. Can control the number of copies of cloned genes
   b. Asexual reproduction
   c. Rapid recombination
   d. Easily cultured
   e. Extensive knowledge of its genetics

8. T_4 DNA ligase is used to do what?
   a. Splice DNA fragments together
   b. Clone DNA fragments
   c. Select for recombinant plasmids
   d. Hydrolyze DNA at specific sequences
   e. Gene vector

Short Answer

9. What are some of the benefits to society from using biotechnology?

10. What are the processes used to clone a gene, and in what order are they used?

Chapter Test Answers

1. **False**

2. **True**

3. **False**

4. **True**

5. **c**

6. **b**

7. **c**

8. **a**

9. Production of pharmaceutical proteins, human growth hormone, more efficient nitrogen utilization by crop plants, human insulin availability, bovine growth hormone, alpha interferon, and increased ethanol production from plant fibers.

10. Isolate plasmid and target DNA, have a suitable vector, digest DNAs with restriction endonucleases, mix DNA fragments together and splice fragments together with T_4 DNA ligase, transform suitable host cells with recombinant DNA, and determine which host cells have the desired plasmid.

Check Your Performance

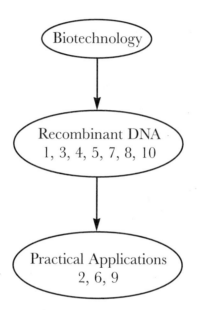

Note the number of questions in each grouping you got wrong on the chapter test. Identify those areas where you need further review and go back to relevant parts of this chapter.

Final Exam

Chapters 1–5 Multiple Choice

1. The ultimate downfall of the theory of spontaneous generation was the result of the work of Louis Pasteur, and required the application of
   - (a) the microscope.
   - (b) Koch's postulates.
   - (c) sterilization.
   - (d) two of the above.

2. Prokaryotes are in the domain(s)
   - (a) Archaea.
   - (b) Bacteria.
   - (c) Eukarya.
   - (d) two of the above.

3. A chemical with a relatively high electronegativity will likely be
   - (a) reduced in a redox reaction.
   - (b) oxidized in a redox reaction.
   - (c) a good reductant.
   - (d) two of the above.

4. The ideal molecule for structural components of the cytoplasmic membrane is a(n) _____ molecule.
   - (a) hydrophilic
   - (b) hydrophobic
   - (c) amphipathic

5. Phosphodiester bonds join
   - (a) amino acids together to form proteins.
   - (b) nucleosides together to form nucleic acids.
   - (c) fatty acids to glycerol to form lipids.

6. S-layers are observed in prokaryotes that are
   - (a) Gram negative.
   - (b) Gram positive.
   - (c) in the Archaea domain.
   - (d) all of the above.

7. The temperature at which a bacterium is grown is _____ the phase transition temperature of its cytoplasmic membrane.
   - (a) greater than
   - (b) less than
   - (c) about the same as

8. The increase in the number of bacteria in a culture growing exponentially is best described mathematically as following a(n) _____ progression, which means that the bacteria double during a constant interval of time.
   - (a) algebraic
   - (b) geometric

(c) numeric

(d) constant

9. Pathogenic organisms are usually _____ and _____.
   (a) barophiles and psychrophiles
   (b) barotolerant alkalophiles
   (c) halophilic neutrophiles
   (d) neutrophilic mesophiles

10. You have been bothered by food spoilage in the crisper drawer of your refrigerator. Most likely, the microbial culprit that has rotted your lettuce is a(n)
    (a) halophile.
    (b) barophile.
    (c) acidophile.
    (d) mesophile.
    (e) psychrophile.

11. In a batch culture of bacteria
    (a) balanced growth occurs only during exponential phase.
    (b) during the lag phase, bacteria are not metabolizing and are not dividing.
    (c) there is no bacterial division during stationary phase.
    (d) the lag phase and the stationary phase are synonymous.

12. Which of the following is (are) necessary for formation of ATP by oxidative phosphorylation?
    (a) A membrane that is impermeable to protons
    (b) An ATP synthase
    (c) An electron transport chain
    (d) An external electron acceptor
    (e) All of the above

True/False

1. The scientific method requires that an hypothesis is testable by experiment or observation.

2. Both endergonic and exergonic reactions may proceed spontaneously.

3. Membranes whose lipids have fully saturated fatty acids will have a higher phase transition temperature than membranes containing lipids with mono-saturated fatty acids.

4. Compatible solutes are used by bacteria for the storage of nutrients and energy.

5. The Pasteur effect describes the decrease in the rate of glucose utilization by facultative anaerobes when they are grown in the presence of oxygen.

Short Answer

1. What property of cells has been used in developing the phylogenetic system consisting of the three domains Archaea, Bacteria, and Eukarya?

2. The _____ bond joins the subunits of proteins, and the _____ bond joins the subunits of nucleic acids.

3. _____ is the energy source for group translocation of glucose by bacteria (e.g., *E. coli*).

4. What should be the reduction potential of a chemical in order that it be an effective external electron acceptor in respiration?

Chapter 6

1. The tricarboxylic acid cycle generates _____ from acetyl-Co-A and oxalosuccinic acid.
   a. *cis*-aconitic acid
   b. *alpha*-keto-glutaric acid
   c. fumaric acid
   d. citric acid

2. Fatty acids break down to _____ as the final molecule.
   a. acetyl-Co-A
   b. propionyl-Co-A
   c. succinyl-Co-A
   d. a and b

Chapter 7

3. Ancestry of organisms is studied by comparing
   a. transfer RNA.
   b. messenger RNA.
   c. ribosomal RNA.
   d. cellular structure.

4. The ninth edition of Bergey's *Manual of Determinative Bacteriology* has _____ sections.
   a. 13
   b. 23
   c. 33
   d. 43

Chapter 8

5. The process that kills the pathogens in milk and juices is called
   a. autoclaving.
   b. tyndallization.
   c. pasteurization.
   d. filtration.

6. The chemical antimicrobial agents that are used topically as skin antiseptics are
   a. alcohols.
   b. glutaraldehyde.
   c. silver nitrate.
   d. a and c.

Chapter 9

7. Assembly of viral components may be done
   a. in the host nucleus.
   b. in the viral nucleus.
   c. outside host cells.
   d. in the endoplasmic reticulum.

8. Prion diseases have these characteristics.
   a. Cause cavitation of neurons
   b. Cause sponge-like appearance in gray matter of brain
   c. Are incurable
   d. All of the above

Chapters 10–11

9. Gene repression is used to regulate the genes coding for the enzymes used in biosynthetic pathways.

10. Transduction and transformation require cell contact between donor and recipient cells before gene exchange occurs.

11. Which of the following characteristics is found in conjugation?
    a. DNase sensitive
    b. Only one gene exchanged at a time
    c. Presence of a plasmid
    d. Passes through a bacteriological filter
    e. All of the preceding
    f. c and d
    g. a, b, and d
    h. None of the preceding

12. Which of the following chromosomal changes can insertion elements induce?
    a. Inversion
    b. Excision
    c. Cointegration
    d. Deletion
    e. Transposition
    f. All of the above
    g. None of the above

Chapter 12

13. Syntrophism is a type of
    a. commensalism.
    b. mutualism.
    c. parasitism.
    d. synergism.

14. Immunization is available against some
    a. endotoxins.

b. enzymes.

c. antiphagocytic factors.

d. exotoxins.

Chapters 13–15 True/False

1. Constitutive host defenses are present and operational always.

2. Self not self-recognition results from the type of antibodies each cell of an organism synthesizes.

3. Antigens are the precursors of antibiotics.

4. An allergen is a foreign substance, not harmful to the host, which triggers the host to produce an inappropriate immune response.

Multiple Choice

5. Which of the following are serological procedures?
   a. Complement fixation
   b. PCR
   c. Restriction endonuclease digestion
   d. ELISA
   e. Immunoelectrophoresis
   f. All of the above
   g. b, c, and d
   h. a, d, and e
   i. a, b, c, and e

6. What feature of antibody molecules gives them a high degree of specificity for epitopes?
   a. Disulfide bonds
   b. A large number of serine and threonine units to form covalent bonds with epitopes
   c. Highly variable domains
   d. Multiple sites of attachment for epitopes
   e. Are synthesized in response to presentation by a particular macrophage

7. Before presenting an epitope to a lymphocyte a presenting cell must
   a. hydrolyze the foreign substance to its basic carbohydrates, amino acids, and nucleotides.
   b. position the foreign substance intact on its surface.
   c. remove all phosphate groups and break all disulfide bonds.
   d. partially digest the foreign substance and position the digested fragments on its surfaces.
   e. all of the above.
   f. none of the above.

8. When an antibody attaches to an epitope on a foreign substance, the antibody may elicit which of the following responses?
   a. Opsinization for phagocytic cells
   b. Release of cytotoxic compounds
   c. Activation of complement proteins
   d. Neutralization of toxins
   e. Blockage of attachment to host cells

f. All of the above

g. a and c

h. b, d, and e

i. b, c, and e

9. The ability of the immune system to produce a stronger response on the second or subsequent exposure to an antigen is called

a. over production.

b. multisensory perception.

c. hyperimmune response.

d. immunological memory.

e. antigenic hypersensitivity.

Short Answer

10. What steps occur between the ingestion of an antigen by a phagocytic cell, such as a neutrophil, and the production of an antibody by a B lymphocyte?

11. Why does a vaccination provide immunity to a foreign substance?

Essay

12. Why can human defenses against foreign antigens be compared to the defense of a city against invaders?

Chapters 16–18 True/False

1. Pathogenicity is the dominant relationship between bacteria and their hosts.

2. Fungi are recognized readily from bacteria and viruses by their larger size, cell walls made of chitin, and frequently filamentous growth form.

3. When traveling, the warning to avoid intestinal infections or intoxications is "boil it, peel it, cook it, or forget it."

4. Prevention of viral infections by vaccination is preferable to treatment because no antibiotics, such as penicillin or streptomycin, are known for curing viral infections.

Multiple Choice

5. Fungus infections are uncommon and mostly superficial in birds and mammals because

a. a nutritional requirement is found only on body surfaces.

b. oxygen deficiency occurs in deeper body tissues.

c. T_C lymphocytes are very effective in destroying fungi.

d. mammals and birds produce funginin, a natural fungicide.

e. none of the above.

6. Bacteria and viruses are transmitted by the airborne route from one human to another most readily by

a. wind.

b. forced air in air handling systems of buildings.

c. dust from dried decaying organic matter.

d. droplet nuclei formed when coughing, sneezing, or vocalizing.

e. contamination picked up as raindrops fall through the air.

7. For a virus to be pathogenic it must
   a. swim between cells on body surfaces.
   b. attach and penetrate cell membranes.
   c. bind essential nutrients to the capsids in its coat.
   d. have an envelope to prevent dehydration.
   e. inactivate tumor necrosis factors and interferons.

8. Which of the following organisms are known to be vectors of certain bacteria or viruses?
   a. Fleas
   b. Ticks
   c. Mosquitoes
   d. Dogs
   e. Mice
   f. All of the above
   g. None of the above

9. Organisms that harbor pathogens during periods of low disease incidence are known as
   a. reservoirs.
   b. vectors.
   c. immunocompromised.
   d. opportunistic.
   e. extinct.

Short Answer

10. What are the major sources of infections of the intestinal tract?

11. Why were HIV and HTLV viruses presented as both wound infections and direct contact infections?

Essay

12. Comment on the following sentences.
    "What me? I don't worry about venereal diseases. I'll just go to the doctor and get a shot of penicillin."

Chapter 19 Multiple Choice

1. The build up of carbon dioxide in the atmosphere, a.k.a. the greenhouse effect, is reduced by
   a. autotrophs.
   b. thermophiles.
   c. acidophiles.
   d. halophiles.

2. Environmental cleanup by microorganisms is called
   a. chelation.
   b. nitrification.

c. bioremediation.

d. biomagnification.

Chapter 20 True/False

1. Recombinant DNA results from joining together DNA fragments from different sources (usually different species).

2. Making different types of DNA is called cloning.

Multiple Choice

3. Which of the following processes or substances are used in the actual process of cloning genes?

a. Restriction endonucleases

b. PCR

c. Plasmid insertion into host cells

d. Promoters

e. Operators

f. All of the above

g. b and c

h. None of the above

4. What can be used to detect which host cells contain a plasmid?

a. DNA ligase

b. Primers

c. Polylinkers

d. Ampicillin resistance

e. Gene guns

5. Bacteria cultured on a nutrient medium in a Petri dish and shown to contain some recombinant DNA are said to constitute a

a. colony.

b. population.

c. genomic library.

d. vector city.

e. none of the above.

Short Answer

6. Once one inserts recombinant DNA into a suitable host cell how are the cells with the desired recombinant gene recognized? How frequently do desired cells occur?

Final Exam Answers
Chapter 1–5 Multiple Choice Answers

1. **c** 2. **d** 3. **a** 4. **c** 5. **b** 6. **d** 7. **a** 8. **b** 9. **d** 10. **e** 11. **a** 12. **e**

Chapters 1–5 True/False Answers

1. **T**   2. **F**   3. **T**   4. **F**   5. **T**

Chapters 1–5 Answers to Short Answer Questions

1. 16S and 18S rRNA

2. peptide :: phosphodiester

3. phosphoenolpyruvate

4. high reduction potential relative to the chemical being oxidized as a source of chemical energy

Chapters 6–12

1. **d**   2. **d**   3. **c**   4. **c**   5. **c**   6. **d**   7. **a**   8. **d**   9. **T**   10. **F**   11. **c**   12. **f**
13. **b**   14. **d**

Chapter 13–15

1. **T**   2. **F**   3. **F**   4. **T**   5. **h**   6. **c**   7. **d**   8. **f**   9. **d**

10. Neutrophil partially digests the antigen and positions the antigenic fragments on its surface with the MHCII molecule. Next the MHCII-antigenic fragment complex binds to a TCR on a T_H lymphocyte and is stabilized by CD 4. In addition B7 on the neutrophil binds to CD 28 on the T_H lymphocyte. IL-1 from the neutrophil binds to IL-1 receptors on the lymphocyte. The T_H lymphocyte releases IL-2, IL-4, and IL-5, which draw B lymphocytes carrying BCR-antifragment complexes. These complexes bind to TCR on the T_H lymphocyte followed by binding of CD 40 on the B lymphocyte to CD 40L on the T_H lymphocyte. Lastly IL-2, IL-4, and IL-5 bind to receptors on the B lymphocyte converting the B lymphocyte into a plasma cell producing antigen-specific antibody.

11. First exposure to an antigen elicits a primary response after about 14 days. Antibodies produced are used or degraded rapidly. With the second exposure a very strong secondary response is activated which lasts much longer and can be duplicated anytime the body is exposed to that particular antigen.

12. In defending a city the defenders construct a series of defenses, one to take the initial charge and others to back up that primary barrier. In the human defense system one sees much the same situation. Skin and mucous membranes serve as the first line of defense. Should that be breached, the phagocytic cells, cytotoxins, and inflammatory reactions of the reticuloendothelial system are activated and then held in final reserve in the specifically inducible antibody-forming system. Apparently this three-layered series of systems is very effective because disease is the exception, not the rule.

Chapter 16–18

1. **F**   2. **T**   3. **T**   4. **T**   5. **c**   6. **d**   7. **b**   8. **f**   9. **a**

10. The four-F rule comes in handy here: food, fingers, feces, flies.

11. These viruses must be delivered into the blood of the uninfected individual from bodily fluids of an infected individual. Direct contact occurs if wounds are made as one is exposed to infected fluids, such as from anal intercourse. On the other hand, the infected bodily fluids may bath preexisting wounds, such as when vaginal intercourse occurs in an individual having lesions from syphilis, gonorrhea, or herpes. Because these viruses can remain infective for some time when deposited on non-living surfaces, such as needles, individuals wounded by those devices become exposed to the viruses.

12. These sentences are very short-sighted and overlook the fact that other individuals may be affected by these diseases.
    a. These diseases are highly contagious for the most part.
    b. Not all of these diseases are susceptible to penicillin therapy.
    c. Some of the physiological effects are serious and irreversible.
    d. There may be transplacental passage of some of the pathogens to infect fetuses or an infant may be infected during birth.

Chapter 19

1. **a**   2. **c**

Chapter 20

1. **T**   2. **F**   3. **g**   4. **d**   5. **c**

6. Success in getting the vector into a host cell depends on the vector and the environmental conditions used during the insertion process. The example given in the text usually yields 1/1,000 to 1/10,000 host cells obtaining recombinant DNA. There are other procedures where efficiency approaches 1/100 host cells.

   First, cells with plasmids are located by incorporating an antibiotic resistance gene or gene that directs the synthesis of a fluorescent pigment. If the polylinker where the recombinant DNA is inserted into the plasmid vector inactivates (knocks-out) a gene, such as an enzyme like β-galactosidase, then one would know that colonies or cells that survive and fail to produce the enzyme are recombinant ones. Then, one must detect the product of the gene of interest in order to find the desired recombinant host cell.

INDEX

197–198, 202, 215, 218, 223, 242, 249, 250
Toxin production, 103
Toxin neutralization, 180
Toxin, botulism, 165, 218
Toxoid, 215, 223
Toxoplasmosis, 249
Transcription, 114, 118, 121, 123
Transduction, 130, 138, 139, 140
Transformation, 130, 136, 137
Transmissible mink encephalopathy, 107
Transmission of disease, 161, 214, 215
Transovarian passage, 222
Transposase, 131
Transposition, 131
Transposon, 91, 130, 131
Treponema pallidum, 126, 208, 220, 221
Tricarboxylic acid, 64, 67
Tricarboxylic acid cycle, 51, 62, 64, 68, 71, 74
tRNA, 115
Tuberculosis, 142, 215, 246
Tumor antigens, 182

Tumors, 233
Turbidostat, 43, 45
Tyndallization, 33, 92
Typhoid fever, 218, 225

Ultraviolet radiation (UV), 88, 92, 132
Unit membrane, 28
Urea, 106
Ureaplasma, 24
Urethra, 170

Vaccination, 160, 184, 215, 216, 223, 228, 229, 233
Vaccine, 191, 229, 231, 234, 236
Vaccinia, 206
Varicella, 228
Variola, 206
Vascular permeability, 173
Vector, insect, 162
Vegetative cells, 31, 32, 33
Viability, 38, 39, 43, 44, 45
Vibrio parahemolyticus, 24, 160, 218
Viral pesticides, 108
Viral host specificity, 108
Viremia, 231

Viricidal/viristatis, 88
Virinos, 106
Virion, 99, 139
Viroid, 106, 113
Virulence factors, 155, 159, 160
Virus, 4, 5, 6, 90, 99–102–105, 113, 139, 156, 197, 206, 215, 228–236, 243, 266
Virus protein coat, 106, 139, 140, 191
Voges-Proskauer, 79

Waste product, 16, 44, 54, 56, 57
Water activity, 40, 41
Water treatment, 94
Western blot, 266
Whooping cough, 184, 216

Xerophiles/xerotolerant, 41, 258

Yeast, 10, 53, 158, 242, 246
Yellow fever, 236
Yersinia pestis, 158, 160, 223

Zone of inhibition, 91
Zoonosis, 162